ANGEL BLUE

by A. J. Butcher

Somewhere in the not too distant future, a teenaged spy is about to go to work.

Lori Angel is a graduate of the Deveraux College, the highly secret school for teenage spies. Until now she's been posted on the coast, doing a whole lot of surfing and partying. But then a fellow graduate turns up at her apartment. Dead. Stashed in his rucksack are a series of clues that will lead Lori into the most dangerous situation of her life. And into conflict with a secret society, every member of which is supposed to be dead.

LORI ANGEL is the new spy novel from the author of the brilliant EDWARD RED. It's an everyday story of rad-surfing, cool parties and body-swapping megalomaniacs trying to destroy humanity.

www.atombooks.co.uk

Spy High Competition

Be a villain or a hero in a Spy High novel!

Who would you rather be? An evil genius intent on world domination or a secret agent who knows just which wire to cut to save the planet at the eleventh hour? You can create your own character – based on yourself of course – and you might get to feature in one of A. J. Butcher's explosive SPY HIGH novels!

Let us know, in no more than a hundred words (and a picture if you'd like), why your character would be the coolest addition to Spy High and you could win the chance to be immortalised in the Spy High hall of fame and £500 worth of Atom Books for your school library.

Send your entries to: EDWARD RED competition, Atom Books, Brettenham House, London WC2E 7EN or email atom.uk@twbg.co.uk. The closing date is 3rd September 2004.

For more information and terms and conditions, visit www.spyhigh.co.uk

QUALITY IMPROVEMENT TECHNIQUES IN CONSTRUCTION

STEVEN MCCABE

LONGMAN

The CHARTERED
INSTITUTE OF
BUILDING

Addison Wesley Longman Limited
Edinburgh Gate
Harlow, Essex CM20 2JE, England
and Associated Companies throughout the world.

Co-published with The Chartered Institute of Building through
Englemere Limited
The White House, Englemere, Kings Ride, Ascot
Berkshire SL5 8BJ, England

First published 1988

ISBN 0 582 30776-7

British Library Cataloguing-in-Publication Data

A catalogue record for this book is
available from the British Library

Set by 35 in 10/12pt Ehrhardt
Produced through Longman Malaysia, CLP

CONTENTS

Foreword		vii
Preface		ix
Acknowledgements		x
1	OVERVIEW	1
	1.1 Introduction	1
	1.2 Objectives	1
	1.3 General structure	2
	1.4 Chapter outlines	2
2	WHAT IS QUALITY MANAGEMENT?	6
	Objectives	6
	2.1 Context	6
	2.2 Getting to grips	6
	2.3 Some definitions	7
	2.4 The four stages	8
	Summary	14
	Questions	14
	Further reading	15
3	THE HISTORICAL DEVELOPMENT OF QUALITY MANAGEMENT	16
	Objectives	16
	3.1 Context	16
	3.2 The origins of inspection and quality control	17
	3.3 Formal QA: the importance of Hawthorne	19
	3.4 A new approach to quality management	21
	3.5 Japan's influence on quality management	22
	3.6 What is Japanese management?	23
	Summary	25
	Questions	26
	Further reading	26

4 THE GURUS OF QUALITY MANAGEMENT 27
  Objectives 27
  4.1 Context 27
  4.2 What is a guru? 28
  4.3 The influence of Japan? 29
  4.4 The pioneers 29
  4.5 The followers 37
  4.6 The contemporaries 43
  Summary 47
  Questions 48
  Further reading 48

5 THE USE OF QUALITY ASSURANCE 49
  Objectives 49
  5.1 Context 49
  5.2 What is ISO 9000? 49
  5.3 Third-party assessment 50
  5.4 The clauses of ISO 9000 51
  Summary 61
  Questions 62
  Further reading 62

6 THE IMPORTANCE OF QUALITY MANAGERS 63
  Objectives 63
  6.1 Context 63
  6.2 Starting on the right foot 64
  6.3 The role of the quality manager 64
  6.4 How not to implement QA 66
  6.5 Writing procedures interactively 67
  6.6 Maintaining momentum 68
  6.7 Other duties of the quality manager 69
  Summary 77
  Questions 77
  Further reading 78

7 FROM QA TO TQM: ACHIEVING A CHANGE IN CULTURE 79
  Objectives 79
  7.1 Context 79
  7.2 Building on QA 79
  7.3 The road to TQM 80
  7.4 Changing the culture 81
  7.5 Can organisational culture be described? 82
  7.6 How to facilitate cultural change 83
  7.7 Planned and emergent approaches to change 84
  Summary 87
  Questions 88
  Further reading 88

8 THE IMPORTANCE OF PEOPLE 89
    Objectives 89
8.1 Context 89
8.2 A review of motivation theory 90
8.3 More contemporary theories of motivation 96
8.4 What is empowerment? 98
8.5 What do gurus say on empowerment? 98
8.6 How can employees be empowered? 100
8.7 Organisational changes to facilitate empowerment 102
    Summary 102
    Questions 103
    Further reading 103

9 LEADERSHIP FOR TOTAL QUALITY MANAGEMENT 104
    Objectives 104
9.1 Context 104
9.2 Leadership and TQM 104
9.3 The meaning of leadership 105
9.4 Different approaches to leadership 116
9.5 Charismatic leadership 117
9.6 Transformational leadership 119
    Summary 120
    Questions 120
    Further reading 121

10 USING QUALITY IMPROVEMENT TECHNIQUES 122
    Objectives 122
10.1 Context 122
10.2 Tools for TQM 122
10.3 Teamwork for TQM 133
10.4 Setting up a team 136
10.5 Stages in team development 136
10.6 Roles within a team 137
    Summary 139
    Questions 139
    Further reading 140

11 OTHER WAYS TO ACHIEVE IMPROVEMENT 141
    Objectives 141
11.1 Context 141
11.2 Partnering 142
11.3 Benchmarking 148
11.4 Business process re-engineering 154
    Summary 159
    Questions 159
    Further reading 160

12 LOOKING TO THE FUTURE: IS THERE LIFE AFTER
    TQM?   161
        Objectives   161
    12.1 Context   161
    12.2 The learning organisation   164
    12.3 Lean production and management   169
    12.4 Becoming world class: prizes and awards   175
    12.5 Looking to the future   181
        Summary   182
        Questions   182
        Further reading   183

13 QUALITY MANAGEMENT IN THE CONSTRUCTION
    INDUSTRY: PRACTICAL APPLICATIONS   184
        Objectives   184
    13.1 Context: Bill's story   184
    13.2 Introducing QA to construction   185
    13.3 My research objectives   187
    13.4 The firms in my sample   188
    13.5 Old faces for a new job   188
    13.6 The quality manager as enforcer   191
    13.7 Writing procedures interactively   191
    13.8 The quality manager as charismatic   193
    13.9 Getting the subbies involved   193
    13.10 Senior managers and QA   197
    13.11 Maintaining momentum   198
        Summary   200
        Questions   201
        Further reading   201

14 THE JOURNEY TO TQM: CONTRACTORS' EXPERIENCES   203
        Objectives   203
    14.1 Context   203
    14.2 A concrete story   204
    14.3 Setting up a quality circle onsite   205
    14.4 Getting close to the client   208
    14.5 Using subcontractors' expertise   209
    14.6 Senior management and TQM   211
    14.7 Cultural change in construction   213
    14.8 Some concluding thoughts   214
        Summary   216
        Questions   216
        Further reading   217

    References   218
    Index   226

FOREWORD

The Construction Industry is the hub for wealth creation, progress and advancement in any nation. It is therefore not surprising to see that the real measure of economic strength has always been 'how thriving' the construction industry sector is.

If one is to look at the major prerequisites for successful performance with TQM, it will be found to include the following:

- Leadership
- Quality Assurance and Systems
- Management by facts and the use of quality tools and techniques
- People Management
- Process Management
- Customer-Focus

These factors are very much ones that drive the construction industry and any other industrial concern to the same extent. Steven McCabe has been very elaborate in highlighting all of the aforementioned factors and many more.

Although there is a proliferation in the number of texts on TQM, *Quality Improvement Techniques in Construction* presents a perspective which is very much applicable to the construction industry, thus filling a gap. It should, therefore, be considered as a welcome addition. Chapters 13 and 14 for example are very original and extremely informative.

This book is well structured and very rich in content. The chapters are set with specific learning objectives, a summary, list of questions to be considered and a reading list at the end. It will be the ideal companion for The Chartered Institute of Building students and other students taking quality and related degrees.

Professor Mohamed Zairi
SABIC Chair in Best Practice Management
University of Bradford

PREFACE

This book aims to meet Levels 2 and 3 in the Educational Framework of the Chartered Institute of Building. It is intended for all students of the built environment.

Quality management has received considerable attention in recent years. People have debated whether techniques developed in manufacturing may be applied to construction. I describe what these techniques are, and how they can be implemented, regardless of context. I believe it is fallacious to treat construction as unique or different. Construction organisations can improve by understanding the appropriate aspects of quality management.

Besides theories, I describe some construction applications to show what can be achieved with quality improvement techniques.

ACKNOWLEDGEMENTS

I would like to thank all those who have assisted me in writing this book.
In particular, I am indebted to the quality managers who extended time and understanding in my research of how quality management is really carried out in the construction industry.

I also wish to thank my wife, Graine, and sons James and John. Their support and encouragement has been much valued. It is to them that I dedicate this book.

Special thanks are extended to James Newall, Rose Artuso, Paul Hawksworth and their colleagues at Addison Wesley Longman.

I am also grateful to all those who gave permission to reproduce copyright material. Whilst every effort has been made to trace the owners of the copyright material in a few cases this has proved impossible and I take this opportunity to offer my apologies to any copyright holders whose rights I may have unwittingly infringed.

OVERVIEW

1.1 QUALITY MANAGEMENT: JUST ANOTHER FAD?

Price makes a point about quality when he states that the 'the word itself must have been used more in the last ten years than in the preceding centuries, yet the more we hear it, the more confusing its meaning seems to become' (1990:3). The word does seem to have become ubiquitous. Everything we do, everything we consume, the service we receive must have quality. But what exactly does *quality* mean?

It is debatable whether there is an answer. Quality, like beauty, appears to be in the eyes of the beholder. What constitutes quality to one person, may be different to others. In an increasingly competitive market, consumer power is recognised as being a major determinant on the ability of an organisation to sell its goods. Thus, the capability to provide what the customer wants is the minimum expectation. Indeed, as subsequent chapters will describe, the desire to give customers more than their expectations is what characterises organisations as excellent. Japan can claim considerable influence in having created the situation whereby quality is now the number one objective for many organisations. Quality is not just another fad; it is here to stay.

1.2 OBJECTIVES

The purpose of this book is to introduce the subject of quality. I assume the reader is relatively unacquainted with quality management, so I aim to cover all the major issues, starting from the fundamentals.

Quality management is the umbrella term for approaches to quality. The various methods and techniques will be described in subsequent chapters. The principles of quality management are applicable to every industry, within all organisations and at every level. But applications do vary from one sector to another, and quality management for construction is dealt with in the final last two chapters.

Theories and techniques may be transferred from between industries. This has been recommended in reports by Latham (1994) and by the Charted Institute of Building (CIOB) (1995). Remember that the construction industry is perceived as

1

having serious quality problems (Ball 1988:217; Building Economic Development Council 1987; Harvey and Ashworth 1993:143–44). Quality improvements made by individual construction companies will ultimately benefit the industry as a whole.

1.3   GENERAL STRUCTURE

Fourteen chapters deal with specific elements of quality management. Each begins with a list of objectives to provide a focus for understanding, and each has a context to help place the subject-matter in the wider sphere of quality management.

Each chapter concludes with a summary of the main points, a discussion question and an individual question. These questions are to stimulate reflection and debate about the issues described. I have also included a selection of suggested further reading, and I would strongly recommend consulting some of the items. The breadth of the subject and its continual development mean that regular reading is very important.

1.4   CHAPTER OUTLINES

Chapter 2 explains the common terms associated with quality management. As such, the British Standard BS EN ISO 8402 (1995) *Quality management and quality assurance – vocabulary* is used to provide definitions. There are also descriptions of the four accepted stages to quality management: inspection, quality control, quality assurance and total quality management.

Inspection and quality control, are not proactive approaches. Quality Assurance (QA), and Total Quality Management (TQM) are among the most significant, and much of this book is dedicated to describing them.

Chapter 3 provides a history of quality management. Despite the belief that quality is a recent phenomenon, references go back as far as 2000 BC. However, techniques of quality such as inspection and control developed in response to the Industrial Revolution. In particular, mass production required a reduction in control by workers. Thus retrospective checking by skilled inspectors came into being.

Quality assurance arose from concerns about mass production. It aims to restore control over the standard of work to the worker, and its development appears to have been influenced by the Hawthorne experiments. The two World Wars encouraged the development of standards for armament supply and led to formal QA using quality systems.

Japan had considerable influence on the advent of TQM. But the irony is that Japan's dominance was achieved with the help of Western advice. Two Americans, Deming and Juran, are noteworthy in this respect. Chapter 3 describes how Japanese organisations benefited from their advice, and how the West has tried to emulate them.

Chapter 4 explains what the so-called gurus of quality management have contributed to the subject (particularly TQM). It covers the pioneers, Deming and

Juran, who provided the earliest philosophies of quality management. As such, they can be regarded as having started the revolution; the followers, Crosby, Imai, Feigenbaum, Ishikawa, Ohno and Shingo, who built upon the advice and philosophy provided by Deming and Juran; and the contemporaries, Peters and Waterman, Pascale and Athos, Ouchi, Conway and Schonberger, who provide theories and advice to organisations facing the challenges of the 1990s.

Chapter 5 explains what QA actually involves and how it can be used by any organisation to implement formal quality. The contents of British Standard ISO 9000, which provides guidance for the implementation of a quality management system, are described in detail. It gives practical examples of how the clauses of ISO 9000 can be applied to construction.

Chapter 6 elaborates on Chapter 5; it contains practical advice on how and, more important, who is responsible for the effective implementation of a quality system. It describes the role of a quality manager. Technical and administrative abilities, to interpret the organisation's business needs in accordance with ISO 9000, need to be coupled with skilful human relations management, crucial when enabling system users to understand and support an initiative.

Chapter 7 describes what comes after QA, and explains how to begin making the transition to TQM. The most important requirement is a change in the culture, to facilitate the transition.

Chapter 7 continues by examining culture and how it applies to organisations. It also explains what conditions are necessary to achieve cultural change. In particular, it compares the paradigms (models) of planned and emergent change, how they relate to cultural change and whether they are appropriate.

Chapter 8 deals with a crucial issue for TQM and cultural change: how to involve the members of an organisation. People's support and commitment is usually achieved by motivation. After reviewing the established theories, Chapter 8 offers useful advice for initiating organisational change using quality management.

It also explains the more contemporary theories of motivation, particularly those models and philosophies which place strong reliance on involving people at every level, especially the operational level. Operators then become more autonomous and are given more control over what they do, often known as empowerment. Empowerment is defined in various ways, including quotations from gurus, and the chapter ends with some suggestions on how to encourage it.

Chapter 9 looks at who will provide the leadership and who will teach the necessary skills. It explains why leadership is important and analyses the relevance of existing theories when applied to TQM.

Contemporary theories of leadership suggest the adoption of different styles. *Charismatic leadership* and especially *transformational leadership* are advocated by those who believe the traditional approaches have limited value.

Chapter 10 explains the tools and techniques of quality management, both hard and soft. Hard techniques include statistical process control (SPC), Pareto analysis and brainstorming. They tend to have a definite application in organisations or teams and they are used to solve problems or to help with the implementation of quality improvements.

Soft techniques attempt to harness the skills and expertise of people in the organisation. They do employ hard techniques, but the emphasis is on getting people to collaborate and cooperate as a group. Teamworking is an essential part of TQM, so Chapter 10 considers how it can be encouraged and what benefits may accrue.

Chapter 11 describe some complementary methods; they are not strictly part of TQM but they reinforce the potential for benefit. Three methods are covered. Partnering relates to the belief that improvement will be more likely if closer relationships are formed both upwards, i.e. with customers, and downwards, i.e. with suppliers and subcontractors. Partnering is an established concept in other industries, and has been used extensively in Japan by car producers such as Toyota. Benchmarking is closely associated with improvement. It is based on the principle that before an organisation can improve, it needs to establish how good it is relative to other organisations, particularly those regarded as excellent. By comparing key business processes, it is possible to establish critical success factors for achieving improvement. The aim is to use benchmarking to improve an organisation until it can compete against the best in its market. Business Process Re-engineering (BPR) considers what an organisation needs to achieve to fulfil a customer's expectations. This is not in the traditional sense of departmental functions, but in terms of the key processes which will enable the customer to get what they want with minimum inefficiency. Any activity which is not essential, or does not add value, should be re-engineered so that it does, otherwise it should be discontinued.

Chapter 12 looks to the future and describes three methods for improvement beyond TQM. The learning organisation encourages an environment where the members are encouraged to be innovative and creative as part of the overall effort to improve. Everybody can reflect on what they have achieved and how they have achieved it. Reflection will lead to understanding and perhaps to suggestions for improvement. When carried out in teams, there will be a consensus for implementing or rejecting the suggestions. The principles of lean production and management were developed by Toyota as a method to reduce and eventually eliminate *muda*, a Japanese word for waste. By using technology in harmony with humans there is vast potential for identifying wasteful processes. Toyota identified certain key concepts as essential to lean production: just-in-time, automation, flexibility in the workforce to cope with change, and the encouragement of creative thinking by the workforce. These principles can also be applied outside the manufacturing environment. World class companies may be distinguished by quality prizes such as the European Quality Awards. Research carried out by Lascelles and Dale (1993:285–96) has indicated there are six levels for adoption of TQM. However, it is only when organisations consider themselves capable of entering for quality prizes that they will be approaching world class status. Chapter 12 looks at how to win these quality prizes.

Chapter 13 describes the actual experiences of several construction organisations when implementing quality management using ISO 9000. These experiences were recorded during my research, using a qualitative methodology (participant observation), into how quality management is being used by contractors. In particular, the emphasis has been on the work of the quality manager: What do they do? How do the do it? And what are the difficulties they must overcome?

According to my research, quality managers identify several issues as crucial. All expressed their belief that quality is neither part-time nor short-term. It requires commitment and dedication, from themselves as those directly responsible for the quality system, the workforce, and especially senior managers, who can sometimes be ambivalent towards QA. Benefits come when QA is implemented 'correctly'. It has helped the quality managers to develop problem-solving approaches and encouraged them to consider other methods for producing organisational improvement.

Chapter 14 describes the experiences of construction organisations which have made, or are making, the transition towards TQM and other improvement initiatives. The traditional culture that exists in the industry does not make this easy. Quality managers who are trying to manage the change have found that improvement is only possible when all those involved have a desire to take part. A transformation in hearts and minds occurs by involving every person who contributes to the process, consistent with the paradigm in Chapter 7.

The quality managers I interviewed agree that a top–down approach is likely to be at best tolerated, and probably resented as managerial interference. They recommend that a bottom–up approach is more likely to bring about the cultural change so crucial to the aspiration of continuous long-term improvement.

Chapter 14 concludes with the advice of a CIOB report (Chartered Institute of Building 1995) which followed a sponsored study of the Japanese construction industry. It describes the radical improvement that Japanese construction has been able to achieve by adopting the established principles of improvement from other industries. It advocates that British construction, although working under very different conditions, should similarly learn that practitioners, together with academia can utilise R&D (research and development) to encourage real improvement. As such, the construction industry can benefit from all the theories and models presented in this book.

CHAPTER 2

WHAT IS QUALITY MANAGEMENT?

OBJECTIVES

- Understand the concept of managing quality.
- Appreciate how approaches differ.
- Distinguish QA and TQM.
- Understand the advantages and disadvantages.

2.1 CONTEXT

Ironic as it may seem, an apparent problem with quality management is the word *quality* itself. For some it signifies excellence, and they may believe that ISO 9000 will automatically make their products or services better than a competitor's. Perhaps it can, but simply having 'quality', as if it were a bolt-on attachment, will not achieve the excellence they imagine. Improvement, which should be the objective, takes more than implementing a system.

Even the British Standards Institution (BSI) could be seen as encouraging these misconceptions. Consider the claims in one of their ISO 9000 advertisements, taken from the *Financial Times* of 15 March 1996:

It's universally recognised . . .
It improves productivity . . .
It gives you a competitive edge . . .
It pays for itself . . .

The introduction of quality management is certainly a phenomenon that has occurred in the last decade or so. But what does it mean? Also what does it seek to achieve?

2.2 GETTING TO GRIPS

When offering a service or product, the minimum objective of any organisation or individual is to provide what is expected. There will be a provider and a customer.

Both are free to obtain what they can from the transaction and there is usually a market consisting of customers who can willingly purchase wherever they like. Providing they have choice, customers will normally seck to maximise their purchasing power. We all tend to buy on a daily basis, and normally we do so on the basis of the maxim: You pay your money and you take your choice. There is a tendency to think that quality has a price. Indeed it may, and to use the oft quoted cliché: You pay Rolls-Royce money and get a Rolls-Royce product.

However, a purchaser has an expectation about what they desire in seeking to maximise their purchasing power. The calculation that occurs is individual and perceptive. In effect, consumers make sophisticated judgements about value. Thus the amount they are prepared to spend provides a guide to the expectations. Normally we do it every day of our lives in order to achieve satisfaction.

Economists call this the theory of utility. It tries to provide an explanation of the decisions we make in terms of apportioning fixed income. We will try to ensure that what we receive will represent good investment.

Reputation plays its part. If you are recommended to use a particular supplier, it is usually because others who have used them think they are good. This does not mean that such suppliers are necessarily expensive. Far from it, they may be extremely cheap in comparison to others. The difference is that their product or service performs well in comparison to competitors. What is worth asking is how do good suppliers achieve their reputation?

Most potential buyers rarely bother to verify how suppliers actually manage their organisation. As long as buyers continue to get what they expect, they will normally continue to purchase from them. But the important point is that being able to supply what customers want is not something which can be left to chance. It requires management, and quality management is the process that any sensible organisation will use in order to consistently satisfy its customers' expectations. It need not be complicated. In fact, it can be summarised as being good or sensible management.

2.3 SOME DEFINITIONS

There are standards dedicated to providing definitions for quality management: BS EN ISO 8402 (1995) *Quality management and quality assurance – vocabulary* and BS 4778 Part 2 (1991) *Quality concepts and related definitions*. They provide an authoritative source for the terms that are used in quality management, but even these terms are often subject to reinterpretation! The definition for quality management is according to BS EN ISO 8402:

> All activities of the overall management function that determine the quality policy, objectives and responsibilities, and implement them by means such as quality planning, quality control, quality assurance, and quality improvement within the quality system. (1995:24)

This suggests there are different approaches to achieving quality management and they are described in the next section.

2.4   THE FOUR STAGES

Dale, Boaden and Lascelles (1994) are typical of those who believe there are four stages of quality management (QM): inspection, quality control (QC), quality assurance (QA) and total quality management (TQM). Figure 2.1 shows the progression from one stage to the next.

Inspection and QC are retrospective; they operate in a detection mode, aiming to find problems that have occurred. QA and especially TQM aim to reduce and ultimately to avoid problems occurring. This means they can be used to bring about improvement. QA and TQM are two of the main approaches covered in this book.

2.4.1   Inspection

BS EN ISO 8402 defines inspection as 'activity such as measuring, examining, testing or gauging one or more characteristics of an entity and comparing these results with specified requirements in order to establish whether conformity is achieved for each characteristic' (1995:22).

Using inspection to ensure conformance is still widely used in some industries, particularly construction. Much of what is built will be compared to the drawings and specifications. Unless the customer agrees otherwise, the contract requires that anything which does not conform will need to be done again until the client is satisfied that it meets the specification.

2.4.2   Quality control (QC)

This stage is often regarded as an extension of inspection. According to BS EN ISO 8402, it involves the 'operational techniques and activities that are used to fulfil requirements for quality' (1995:25). QC will require collection of data in order to use statistical techniques. From this information, trends will often emerge which show where certain problems are occurring. This technique is used as a matter of course in manufacturing. It is much rarer in construction. Statistical analysis of concrete cube test results is the one situation where statistics are routinely used.

2.4.3   Quality assurance

Dale, Boaden and Lascelles make the following statement: 'In a detection or "fire-fighting" environment, the emphasis is on the product, procedures and or service deliverables and the downstream producing and delivery processes' (1994:6). As they explain, considerable effort normally goes into removing faults or problems before the product or service reaches the customer. However, this is not satisfactory because 'in this approach, there is a lack of creative and systematic work activities, and planning and improvements are neglected. . . . Problems in the process are not removed but contained' (ibid.:6). As they stress: 'An environment in which the emphasis is on making good the non–conformance rather than preventing it arising is not ideal for engendering team spirit, co–operation and a good working climate.

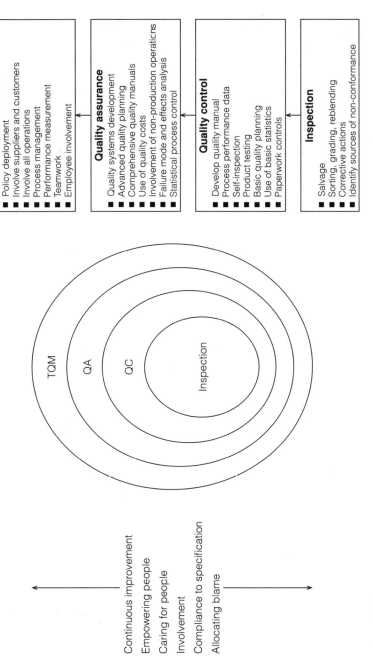

Total Quality Management
- Policy deployment
- Involve suppliers and customers
- Involve all operations
- Process management
- Performance measurement
- Teamwork
- Employee involvement

Quality assurance
- Quality systems development
- Advanced quality planning
- Comprehensive quality manuals
- Use of quality costs
- Involvement of non-production operations
- Failure mode and effects analysis
- Statistical process control

Quality control
- Develop quality manual
- Process performance data
- Self-inspection
- Product testing
- Basic quality planning
- Use of basic statistics
- Paperwork controls

Inspection
- Salvage
- Sorting, grading, reblending
- Corrective actions
- Identify sources of non-conformance

TQM

QA

QC

Inspection

Continuous improvement
Empowering people
Caring for people
Involvement
Compliance to specification
Allocating blame

Fig. 2.1 The four stages of quality management (Reprinted with the permission of Prentice Hall Europe, adapted from Dale, Boaden and Lascelles 1994:5)

The focus tends to be on switching the blame to others, people making themselves "fireproof", not being prepared to accept responsibility and ownership, and taking disciplinary action against people who make mistakes' (ibid.:7–8).

Chapter 5 goes into greater detail, but essentially what is being advocated is that any organisation should aim to logically 'prevent rather than cure' problems. In effect, using quality management should be proactive rather than reactive. This is what QA seeks to achieve by using a recognised quality management system such as ISO 9000.

BS EN ISO defines quality assurance as being 'all the planned activities implemented within the quality system, and demonstrated as needed, to provide adequate confidence that an entity will fulfil requirements for quality' (1995:25).

QA using ISO 9000 is covered in more detail in Chapters 5 and 6. But right now it is useful to consider why organisations should implement QA. What do they expect to achieve? Note the belief that QA has two sides, good and bad. There may be benefits, but there may also be pitfalls.

Benefits

Meeting customer requirements   Many consider QA to be synonymous with the use of ISO 9000. This is not always the case. Experience may suggest it is easier to deal with a single person. Frequently this is because the person you deal with also does the work. If you tell them what you want, and providing they have understood, the result will usually be what was expected. The potential for a communications breakdown is somewhat lessened. And remember that small organisations are normally highly dependent on repeat custom. If they do a poor job, you will not use them again. Even worse, you may tell others of the problems you experienced.

It is not uncommon for small traders to use something simple like a notebook to take a customer's requirements. If they are unsure of something, they will ask. In this sort of situation, it is also probable that they will carry out the work. Not surprisingly, such a person will concentrate their effort on achieving customer satisfaction.

But in many organisations, particularly the larger ones, it is not possible for the person dealing with the customer to carry out the work. Even if they do, there will probably be many customers at the same time. This situation requires more advanced management to achieve customer satisfaction. Quality assurance should provide some help.

Communicating customer requirements   The word *communication* is at the heart of quality management. It may be assumed that all organisations, large or small, wish to satisfy their customers, so they need a full understanding of each customer's requirements. How do you begin to achieve a requirement without knowing what it is? All that a quality system seeks to achieve is to ensure that customers get what they want.

The sophistication of a QA system will depend on the complexity of the end result and the number of people involved. For the small one-person firm, it may be informal and relatively undocumented. If the person consistently meets customer expectations, then that is all that counts. They are operating something worthwhile, a system to manage quality that suits its objective.

Staying on tender lists and getting new business   It is widely accepted that, for many organisations, the initial reason for implementing a formal system of QA such as ISO 9000 was due to client demand. Large clients, such as government departments, insisted their suppliers should be accredited as a minimum criterion for consideration. This is a good reason for getting QA. Without it an organisation risks losing business.

Some might argue that having ISO 9000 in order to stay on tender lists is the only reason for implementing it. Although cynical, perhaps, it is true in many organisations. However, advocates of QA believe that organisations who implement ISO 9000 in the right way will have demonstrated that they take quality seriously. For some customers this commitment may justify their use of an accredited organisation in preference to an unaccredited organisation. It is not something which can be guaranteed, but more likely than not, QA accreditation indicates a business that satisfies customers.

Doing it right first time   This may be regarded as a cliché when used about QA. It will be a strange organisation that is not trying to do things right first time. Doing things wrong will cost money, and worse, it may lose customers.

The procedures needed to govern work should be based on the simple principle of ensuring the task will achieve the correct end result. The aim is that every part of the process is a part of the whole. Tasks or operations will need to be right, so they can fit in with other elements or subsequent operations. Construction planning usually works on the need for one trade to complete at a particular time; this allows several trades to commence simultaneously (known as a burst). If a trade has not been done right and therefore has to be rectified, it will probably delay the whole programme. The critical path may be extended, and the job will be handed over late. This will normally cause liquidated damages, unlikely to please anyone. Alternatively, the contractor may have to spend extra money to get back on programme.

It may be argued that the construction contract does at least protect all the parties. Nevertheless, it seems to me that delays or overspends are all too frequent. The time and expense incurred by mistakes is a burden on all involved. It is more sensible to avoid them, and QA may help to do this.

A formal and documented approach may have other advantages:

1. It provides a guidebook to new recruits in an expanding organisation.
2. Teamwork increases because all parties must come together at the start of a job.
3. It identifies areas in which staff require training.
4. Discipline increases when audits are anticipated.
5. Outside bodies may verify that a quality management system is being operated.
6. Appropriate records document what was carried out when, how and by whom.

Pitfalls

Bureaucracy   Increased paperwork is the most frequently cited disadvantage. This is usually because procedures need to be written and documented, leading to bureaucracy. Every industry that has implemented QA appears to suffer from bureaucracy, but it seems particularly acute in the construction sector. This may be because, until recently, the culture of construction considered it more important to do the work than to complete the paperwork.

Cost   There is cost involved in setting up a quality system. Initially there is the need for time and expense to be dedicated to the task of achieving registration. There will be the cost of someone to manage the system. This is because ISO 9000 requires a management representative to be appointed (Chapter 5). There will be the continued expense of regular audits by the third-party accreditor. All of this expense must be allowed for. Unfortunately, construction is an industry which operates on low profits, often with fewer employees than would be desirable. Because of cost, QA in most industries, particularly the construction industry, has not been a welcome addition. It is often said, 'We cannot afford to do it.' But so-called excellent and world class organisations have demonstrated that you cannot afford not to do it.

Using ISO 9000 on its own will not be sufficient to emulate the world class organisations. More will be required. But it is a very good foundation for more advanced methods of QM.

2.4.4   Total quality management (TQM)

Although it is not a prerequisite, TQM often follows the implementation of QA. This is a normal transition and should not be interpreted as QA having failed.

The change form QA to TQM will need to be carefully managed. Although it normally requires the use of procedures and may be criticised for being too formal, QA does have the advantage of being tangible. It is possible to see how well the system is being accepted by auditing. If the procedures are being adhered to, the QA system can be judged successful. If procedures are not being adhered to, then they need to be rewritten or the users need more explanation of what is required.

TQM is less formal, having neither system nor procedures, but its very lack of formality makes it more difficult to describe. It is often described as a philosophy, which requires change in things like attitude, management style and culture. According to BS EN ISO 8402, TQM is a

> management approach of an organization, centred on quality, based on the participation of all members and aiming at long-term success through customer satisfaction, and benefits to all members of the organization and to society. (1995:27)

Note that two British Standards are dedicated to TQM, BS 7850 Part 1 (1992) *Guide to management principles* and BS 7850 Part 2 (1992) *Guide to quality improvement methods.* As their titles indicate, they provide guidance only. Unlike ISO 9000, BS 7850 is not meant to be implemented and assessed by a third party.

Stephen Robbins (1994:51) provides a list of his five essentials for TQM:

- Intense focus on the customer
- Concern for continual improvement
- Improvement in the quality of everything
- Accurate measurement
- Empowerment of employees

It is significant he puts the customer at the top. The word *customer* is not only the end customer as QA tends to imply. In every process there are various stages of production in order to provide the end result which gives the customer what they want. At every stage of the process, one group of people pass on the goods to another group of people. This is like the relationship between traders and buyers. In effect, the receiver of the goods is a customer. They have expectations which, if satisfied, will assist in providing the end customer with what they expect. This concept describes the internal customer, who is crucial to the philosophy of TQM. It is only by addressing all the constituent parts of every process that total improvement can be achieved.

How is TQM applied?

If an organisation is to develop TQM, it must start with its own processes. Dale, Boaden and Lascelles (1994:10) show how effective TQM dictates that QM is 'applied in every branch and at every level in the organization.' As they explain, this may mean that some areas of an organisation need no more than already exists for QA. However, in order to support TQM across the organisation, something beyond simple QA is normally required. There needs to be 'a broadening of outlook and skills and an increase in creative activities from that required at the quality assurance level' (ibid.:10).

TQM is essentially the use of tools (Chapter 9) to provide the accurate measurement cited by Robbins. There will also need to be a greater appreciation of those who are involved in the totality of the process. The aim is ultimately the same as QA. This is to provide customers with what they want. The difference is that potential for improvement in the end result comes from considering every area in which the organisation is involved.

TQM also recognises that people are the key resource in each contributory area. By getting them to be more actively involved, the desired improvement is more likely to occur. This is the concept of empowerment (Chapter 8). The involvement of people is often called the soft aspect of management, compared with the hard aspects of QA rules and QA systems.

Dale, Boaden and Lascelles suggest the concern about 'personal values' resulting from TQM extends to 'include partnerships with suppliers and customers' (ibid.:10). I believe this crucial aspect holds significance for every industry, and a detailed description comes in Chapter 11. The construction industry can learn from those that avoid becoming embroiled in contractual disputes. Learning how others work together to produce a result of mutual benefit should help to lower the cost and raise the chance of satisfaction, an alternative that may work for construction.

SUMMARY

This chapter has described the terms and approaches generically known as quality management. The following aspects are important:

1. Quality is a difficult concept to define in the absence of other measures, but value is something all customers are expected to maximise.

2. Organisations which supply customers will normally seek to maximise the value their customers receive, and quality management will help them to do this.

3. Quality assurance (QA) uses a quality management system, such as the systematic method, to help ensure the processes of an organisation are arranged to deliver exactly what its customers require.

4. QA has great potential to benefit an organisation's competitiveness, but problems may well occur. They can be alleviated by correct implementation of the quality system.

5. Total quality management (TQM) is a less formal approach than QA. It focuses on giving the customer an excellent service and/or product by the use of continuous improvement, employing quality tools and techniques, and by emphasising the importance of people.

6. TQM encourages every person or organisation involved to contribute to the overall effort of improvement. This includes parties outside the direct control of the organisation but which supply it with goods and services.

QUESTIONS

Discussion

Suppose QA is sufficient to meet a particular set of customer requirements. Why would it be adequate merely to meet expectations? Then discuss why TQM is used in other situations to enhance the end result that the customer receives.

Individual

Draw up a report of no more than six pages to explain the following items to practising managers:

(a) How to advise a newly formed company on selecting quality management approaches for the short to medium term and for the long term.

(b) How a new company might avoid the pitfalls of poorly conceived QM initiatives and/or inadequate implementation.

FURTHER READING

Bank, J. (1992) *The essence of total quality management*. Prentice Hall, Hemel Hempstead.

Brown, T. (1993) *Understanding BS 5750 and other quality systems*. Gower, Aldershot.

Schmidt, W. H. and J. P. Finnigan (1993) *TQ manager: a practical guide for managing in a total quality organization*. Jossey-Bass, San Francisco CA.

Wille, E. (1993) *Quality: achieving excellence*. Century Business, London.

CHAPTER 3

THE HISTORICAL DEVELOPMENT OF QUALITY MANAGEMENT

OBJECTIVES

- Explain the historical origins of quality.
- Link the development of quality to management theory.
- Describe the TQM approach.
- Give an outline of Japanese management.
- Describe what the construction industry can learn from Japan.

3.1 CONTEXT

Carol Kennedy makes the point that, for many, quality (in whatever manifestation) appears to be a recent phenomenon:

> The astonishing fact about the quality revolution is how recent it is. In 1980, almost nobody in senior management in Western countries considered quality to be a management issue. Quality was regarded as a matter of inspection and of correcting discovered faults, rather than managing the production process so as to eliminate faults. Today, total quality management (TQM) has a strong claim to be the most influential management theory of our time, with its objectives enshrined in government quality audits such as ISO 9000 and BS 5750. (1994:216)

My research indicates that, for most people in construction, the reason for becoming interested in quality was the introduction of BS 5750. This standard was published by the British Standards Institution for the management of quality. The UK government's desire to promote the standard meant that agencies for capital construction by central government, like the now defunct Property Services Agency (PSA), promoted BS 5750. In effect, many construction organisations felt that, without the standard, they risked losing work.

There was a history behind BS 5750, and its development can be linked to two influences which occurred after the Second World War:

1. In the defence industry of Western Europe and America there was concern at the high rates of failure of equipment in service.
2. Postwar Japan, became pre-eminent in the electrical and automotive sectors (Joynson and Forrester 1995).

But quality dates back much further than that.

3.2 THE ORIGINS OF INSPECTION AND QUALITY CONTROL

The use of measurement is something that has gone on for as long as humans have traded with one another. This enabled agreement so that it was possible for the buyer and seller to agree on what they expected from the transaction. The earliest documented example of the use of quality control concerns King Hammurabi of Babylonia (2123–2081 BC). He ruled over what would be called an early, but relatively advanced civilisation. Many point to the laws that were introduced during his reign as evidence of ensuring compliance with standards. One in particular should be of interest to contemporary students and practitioners of construction. It tends to be quoted more often than any other, perhaps as a warning to those who fail to deliver quality!

> The mason who builds a house which falls down and kills the inmate shall be put to death. (George 1972:9–10)

The development of standards to control the quality of goods in England has been traced back to Saxon times (Drew 1972). It continued ad hoc until the Industrial Revolution, which began in the late eighteenth century. As any student of history will confirm, it is regarded as being of great significance. This is because it caused sweeping changes which forever altered society. It also led to new production methods. The Industrial Revolution also changed the relationship between traders and their customers:

> Even if the principles of trading did not change, emphasis and practice certainly did. Purchasers soon began to demand certainty in quality as never before. Reliance on quality was an important factor. . . . Suppliers were forced to seek what protection they could against redress sought by customers for alleged faulty goods and services. Thus were born comprehensive specifications and contract conditions. . . . From this crude and elementary process emerged the principles of quality assurance as they are understood today. (Taylor and Hosker 1992:6)

In order to achieve increased levels of production, and consistency in standards, merchants embraced the factory system. This required that all the elements of production were organised in one place. The characteristics of this system were

- Complete control of workers
- Division of labour and specialisation
- Long hours of work
- Harsh regimes and discipline, more akin to prisons

Not surprisingly, workers were often unhappy, and labour turnover was high. This caused problems for the factory owners. In particular, they needed to raise wage rates, especially for those workers whose skill was most in demand, not a situation they welcomed. Factory owners demanded the development of technology using machines which required little skill.

As a result of the new technology, many of the entrepreneurs who owned the factories were able to increase production in order to make more money. However, with the increase in work came the realisation they could no longer effectively control everything themselves. There was a need for others to be employed to assist. Thus managers came into being, and despite not owning the factories, they were often given control equal to the owners.

Managers in the early factories spent much time and effort considering how they could obtain increased efficiency from their organisations. The part played by Frederick W. Taylor (1856–1915) was particularly significant. His influence originated what was called scientific management; today it is commonly known as mass production. Scientific management was based on the belief that organisations were like machines. Consequently, they could be operated on scientific principles.

Taylor believed that careful planning and organisation, carried out by managers, would enable manual workers to become more productive. This required the analysis of the operations to be completed. By optimising their sequence, wasteful effort would be reduced. Taylor stressed that managers were to exercise absolute discretion over what the workers did. Managers were also to stipulate standard procedures and times for each job. The advantage of the organisation being more efficient was that more money would be available to pay the workers as an incentive to produce.

Taylor's book, *Principles of scientific management*, published in 1911, laid down the foundations for all subsequent organisational theory. In it he explained there was an alternative to the often erratic methods that had been employed. There was in fact 'one best way' with four essential steps:

1. Managers gather all knowledge traditionally possessed by skilled workers.
2. Managers take the decisions as to how this knowledge is translated into rules or laws to govern the running of the organisation.
3. Every task is governed by these rules, and the person carrying them out should follow instructions without question. The instructions should set out what is to be done, the appropriate method and the exact time allowed.
4. Managers are governed by these rules in exactly the same way as workers.

But others also contributed to the development of early organisation theory: Frank and Lillian Gilbreth, Henri Fayol and Max Weber. All were sympathetic to the principle of scientific management, and most especially that 'planning could be divorced from doing'. However, there was some concern expressed that, because of the monotony and deskilling this system required, the actual quality of the goods being produced would suffer. As a result, a role was created to monitor the standard of goods at the end of production. This task was to be carried out by quality inspectors.

The First World War is significant in relation to quality. The large armaments requirement for the war, which also needed most men of working age to fight, meant that a huge unskilled and inexperienced workforce was suddenly required. This new workforce, primarily women, usually with minimal training, were set to work carrying out simple and repetitive tasks. Any skilled workers who remained became inspectors of the new workers.

Because of the efforts of the workers, remarkable outputs were achieved to assist the war effort. In peacetime, though, many realised the system was neither efficient nor cost-effective. In particular, the belief that quality could be adequately controlled solely by inspection was seriously questioned.

The main criticism was of the belief that any skill could be replaced by routine and procedurised operations. As critics of this system pointed out, all the expertise was located with the managers. Employing inspectors to check on the work of those manual workers operating in this system was not likely to help, because

> inevitably, since they had to be capable of distinguishing between what was satisfactory, and what was not, the inspectors were drawn from the more intelligent and knowledgeable members of the workforce. This had the immediate effect of diluting the skills of the productive team and at the same time creating a powerful group whose existence could be justified only if it found work which had to be rejected. It is difficult to imagine any arrangement more calculated to de-motivate the productive work force and create friction and bad feeling. (Ashford 1989:4)

Studies indicated that workers in this system tended to care very little for what the customer wanted. They merely wanted their work to be passed by the inspector. Often they tried to deceive the inspectors, who in turn retaliated by requiring more inspections and writing more stringent procedures. This would enable them to 'strengthen their stranglehold on their enemies in the production departments' (ibid.:4–5). As Ashford explains:

> An inspector can identify a fault only after it has been committed. He may then order the item to be scrapped or rectified in some way. Whatever the decision, waste will have occurred and harm done which cannot be undone. In many cases the inspector will know the cause of the fault and how it can be prevented, but he has no incentive to pass on this knowledge to those in charge of production. They are on the opposing side and are unlikely to welcome his advice. (ibid.:4–5)

Put simply, if problems are eradicated, where does that leave the inspectors? The answer is obviously jobless, not something that most people want. Hence the critics of scientific management argued, 'as production goes up, quality goes down.' They suggested there had to be a better way. This is the *raison d'être* proposed by advocates of QA.

3.3 FORMAL QA: THE IMPORTANCE OF HAWTHORNE

Concerns about the use of scientific management prompted some research at the Hawthorne factory after the First World War. Its aim was to consider better ways of managing and organising, and it led to what became known as the human relations approach. Its recognition by managers helped to provide an alternative to retrospective inspection and quality control.

The location of the Hawthorne experiments at a plant in Chicago owned by the Western Electric Company in the 1920s and 1930s is significant. As Kennedy remarks, 'Hawthorne in the 1920s was a crucible of influential management research, going down in industrial history as the source of industrial sociology' (1994:217).

The Hawthorne research is so famous because it challenged the notion of using scientific management techniques. But not only that, W. Edwards Deming and Joseph Juran worked at the Hawthorne factory. Because of their efforts in assisting the revival of postwar Japan, they became very influential in quality management. Significantly, Demings' mentor, Dr Walter Shewhart, pioneered statistical process control (SPC) at Hawthorne. Deming was to develop it to great effect during America's war effort in the 1940s, and subsequently in Japan. Despite sounding very technical, the tools it provides are sensitive to people.

3.3.1　What were the Hawthorne experiments?

The original aim of the work carried out at Hawthorne was based on the principles of scientific management. The intention was to enable engineers to establish the effects of lighting levels on worker productivity. Illumination levels were varied for one group of workers and the results were compared against a control group where no changes were made. The assumption was that productivity would alter according to the level of lighting. To the surprise of all concerned, once the experiments started, output increased whether the lighting was turned up or down. Output only decreased when the lighting was reduced to the level of moonlight. Stranger still, the control group's output also increased.

Clearly this caused the Hawthorne engineers to rethink their original hypothesis. But what had occurred? It was at this point that Harvard University was contacted and Elton Mayo became involved. His name is now synonymous with Hawthorne and human relations. Mayo and his teams discovered that the productivity improvements were not caused by working conditions, but because the workers taking part in the study felt special. This was due to the attention they received during the experiments.

Two major findings came from the research and provided the basis for the human relations approach:

1. Informal groups are important because work is seen as a collective and cooperative process.
2. Humans are more than the rational economic beings that Taylor believed; they need security and recognition.

Thus came the rejection of the mechanistic–rational approach that scientific management implied. Managers were advised that workers have emotional needs; they do not respond positively to being controlled with rules and unbending discipline. In particular, Mayo stressed that workers liked to be in groups which provided fraternity and stability, so managers should collaborate with workers to achieve them. Shewhart and Deming passionately believed in this idea.

3.4 A NEW APPROACH TO QUALITY MANAGEMENT

Hawthorne and subsequent research demonstrated how fallacious it was to treat workers as unthinking robots that simply followed instructions. It suggested that workers were capable of making judgements about the standard of their own work, in the same way as skilled artisans before the factory system. Furthermore, the mere fact they could be trusted to make such judgements would enhance their emotional needs. Change, however, did not come rapidly.

Wars seem to play a significant part in the history of quality. The First World War initiated mass production (requiring methods of mass inspection). During the Second World War, Britain and America exchanged their findings on quality management, but neither realised the significance of Dr Deming's work on SPC. Although it had been used very successfully to assist the American war effort in producing armaments, two generations elapsed before SPC was adopted by civilian companies in the United States; and fifty years later, British companies are still learning.

Conditions after the Second World War meant a return to mass production in both America and Britain. Consumers starved of luxury goods were happy to accept them in whatever condition they arrived. But another threat of war provided further impetus for change. Britain and America found a new enemy, the USSR, and began producing new weapons in anticipation of hostilities. But the weapons were produced using scientific principles and they failed at an alarmingly high rate. Government inspectors were employed to cure the problem, but their presence was counterproductive. Ironically, the government inspectors undermined the commitment of internal inspectors, just as the internal inspectors had undermined the workers. When the situation was analysed, it revealed two things:

1. Quality must be controlled over the whole production chain, not left as a retrospective task.
2. Personnel need to be encouraged to accept responsibility for the quality of what they achieve.

Moreover, in May 1968, due to concern at the increasing cost of the failures and accompanying dangers to personnel, the North Atlantic Treaty Organisation (NATO) issued Allied Quality Assurance Publication 1 (AQAP-1): *Quality Control System Requirements for Industry*. It laid down what contractors supplying NATO needed to do in order to satisfy contractual requirements. Significantly, it required contractors to provide documentary evidence that their products conformed. AQAP-2, a guidance document as to how this should be achieved, was published in September of the same year.

The Ministry of Defence in Britain followed the example of NATO by producing documents called Defence Standards (DEF STAN), the UK equivalents of AQAPs.

The first was called 05–08 and was published in March 1970; it was followed by 05–21 in January 1973. These standards signalled the end of *caveat emptor* (buyer beware) and laid the foundations for a commercial standard in QA. It began in 1971, when the Confederation of British Industry (CBI) requested the British Standards Institution (BSI) to investigate how the Defence Standards could be replicated in general manufacturing.

Initially BSI produced two guidance standards: BS 4778 *Glossary of terms used in quality assurance* came out in 1971 and BS 4891 *A guide to quality assurance* came out in 1972. Even BSI had problems with some of the terms they were supposed to be clarifying, interchanging *quality assurance* with *quality control* in BS 4891.

It was another seven years before the publication of BS 5750 *Quality systems*, intended to become the UK Standard for QA. In 1987 it was reissued to correspond with ISO 9000, the international equivalent, and in 1994 it was renamed BS EN ISO 9000. In practice, however, it is usually called ISO 9000. Its contents are fully described in Chapter 5.

3.5   JAPAN'S INFLUENCE ON QUALITY MANAGEMENT

The Second World War had seen the defeat of the two primary aggressors, Germany and Japan. It is now somewhat ironic that their very defeat contributed to a revival in industrial might. In the case of Japan, economic progress has been remarkable, and in quality spectacular. Some have been moved to state that, because of Japan's devotion to quality, they have been able to 'achieve, by peaceful economic means, what they had failed to do by war: to dominate the world' (Morrison 1994:52). How have they been able to do this? By using American expertise.

In the immediate aftermath of the war, General MacArthur controlled Japan. As a form of punishment for their aggression, he dismissed all the senior and middle management from large companies. This act provided the foundations for the Japanese Quality Revolution. Japan had previously been a feudal society. Age was respected and was one of the main criteria for promotion. Suddenly everything changed. MacArthur put in charge younger managers, who until then had only worked in operational areas. They had not been steeped in the traditions of their predecessors, so they were not expected to follow the accepted practices. They were starting from scratch and they could afford to take risks. As Bank states, the result was to 'transform Japan from a largely illiterate, semi-industrialized country turning out cheap and unreliable copies of Western products, into today's nation with the world's highest per capita income and a producer of best quality goods' (1992:60).

Just consider the effect Japan has had on others countries. If you have bought electrical goods recently, it's very likely they are Japanese (though perhaps made here in the UK). Moreover, despite their price having reduced in real terms, they will usually be very reliable and come with a multitude of features. This stems from management methods the Japanese have refined. Competitors have found their traditional goods can no longer compete. They have had to get better or get beaten.

3.5.1   Rover: influenced by Japan?

Many companies have followed the Japanese approach with great success. Rover is a good example of a company that came back from the brink. Its products are now regarded as being as good, if not better than those produced by Japanese competitors. How has this been achieved?

Longbridge in Birmingham, the plant normally associated with Rover has a place in industrial history, but not for the same reasons as Hawthorne. In the 1970s what was then British Leyland experienced often bitter and protracted strikes. The common perception was that the unions were at fault. The plant convenor, Derek Robinson (disparagingly nicknamed Red Robbo), was portrayed in the media as being primarily responsible.

It seemed the company was unmanageable, always on the brink of closure. But closure was unthinkable for a state-run company like British Leyland. Yet in reality the cars it produced were often not very good. In fact, many had severe problems. Buyers increasingly found that foreign models were cheaper and more reliable. Something had to change.

Since then, many things have indeed changed. The company name, the owners, much of the workforce, but most notably the reputation of its product. So much so that in 1992 it was presented with the British Quality Award for what the citation described as 'significant and consistent progress made over the last four years in product quality, efficiency, and people involvement.'

This transformation has not been easy, and one of the things Rover recognised was that they did not have a worker problem, they had a management problem. According to the award citation, they have moved towards 'people involvement'. This required a fundamental change in the way the management conducted themselves in their dealings with workers, something Rover learned by collaborating with Honda.

Now, besides producing cars which are comparable to the best, Rover can provide stable employment. Significantly, workers are much more involved in the process. This is a radical change in culture.

As many now recognise, poor industrial relations tend to be a product of management which does not command respect. This was the case at Longbridge. It is therefore appropriate to explore what Rover and many other organisations have learned from working with the Japanese.

3.6   WHAT IS JAPANESE MANAGEMENT?

In *The road to Nissan*, Peter Wickens describes his experience as a director of personnel and information systems. He concludes there is a 'tripod' of characteristics which he believes are 'by no means unique to Japan but are practised in Japan to a greater extent than anywhere else' (1987:38). The three characteristics are

- Flexibility
- Quality consciousness
- Teamworking

Wickens also believes they can be practised by any organisation. Japanese management also has other characteristics which are believed to provide an environment conducive to cultural change. Many commentators have attempted to provide a definitive set of these features. Burnes (1996) believes they fall into two categories:

- Personnel and industrial issues
- Business and manufacturing practices

The first category, Burnes argues, is the major reason for the success of Japanese companies and it 'comprises a group of practices and policies designed to socialise and bind employees to the organisation, and promote their long-term development and commitment' (ibid.:100). He provides a list that includes the following items:

1. *Lifetime employment.* In Japan it is not uncommon for large organisations to recruit employees straight from school. Such employees frequently remain there for the rest of their working lives.
2. *Internal labour markets.* There is a preference for companies to promote on the basis of performance and avoid bringing in outsiders.
3. *Seniority-based promotion and reward systems.* This follows on from item 2; it means that promotions are given to people who have been with the organisation for a significant amount of time. People do not need to move elsewhere to gain promotion.
4. *Teamwork and bonding.* This is the feature that commentators like Wickens most readily typify as reflecting Japanese management, and which can be practised by any organisation. It means that people regard their co-workers as being 'like an extension of family'.
5. *Single unions.* It is accepted that every person will be part of a union, including senior managers. However, there is only one union for an organisation. They also tend to be more like associations than a Western trade union.
6. *Training and education.* This is another of the key features which many are keen to stress as being crucial to how the Japanese have achieved their success. This book returns to it many times.
7. *Welfarism.* The provision by an organisation of many benefits to employees and their families is widely practised. As a result, the workers feel gratitude and they become reluctant to leave.

What is being provided for employees is an environment which encourages loyalty and commitment. I know an anecdote about a Japanese worker who coped after losing his family in a tragic accident. But after losing his job he felt that 'life was no longer worth living,' so he committed suicide.

Many commentators refer to the importance which Japanese organisations attach to their key resource – people. Often described as a cultural phenomenon, it will be considered in detail in subsequent chapters.

The second category relates to the way the Japanese organise and conduct themselves in business. This is a really crucial issue concerning quality to customers. As Burnes points out, 'The Japanese ability to satisfy customers, and thereby capture markets, by developing and producing products to a higher specification and lower cost than their competitors, is staggering . . .' (1996:102). He believes the following elements characterise their approach:

1. *Long-term planning*. This refers to the fact that because of a different financial infrastructure, Japanese organisations take a longer view when considering investment. I have heard of an organisation which had a business plan for a hundred years.

2. *Timeliness*. The ability to bring out new products, or variations, in a much quicker time than would be possible in the West appears to be a result of just-in-time and lean production. Both aim to reduce cycle times and wasted effort. The Japanese tend to bring together parts of the team from the earliest point.

3. *Dedication to improvement*. This is what many believe is the reason why Japanese products are so good. Indeed the Japanese have a word, *Kaizen*, meaning continuous improvement. This is the core concept of TQM, and according to Gilbert:

> Small steps and often is a better way to produce a controlled evolutionary improvement than large, often unpredictable steps. . . . Kaizen, which can mean, 'transformation to the better' . . . [in] practice . . . means that improvement is a continuous part of the job and demands a different type of management and supervisory thinking. Supervisors become team leaders and educators; managers become enablers providing resources and direction. (1992:164)

As is also suggested by those who advocate emulating Japanese management methods, there must be an environment where so-called hard techniques like SPC are combined with the softer people approaches described above. They should be used so they are 'integrated and mutually supportive of each other' (Burnes 1996:104). The use of such hard and soft tools and techniques will be described in Chapter 10.

SUMMARY

A number of issues have emerged during the historical development of quality management:

1. Quality is not a recent phenomenon.

2. The Industrial Revolution which led to the development of mass production (also called Taylorism), meant that skilled workers were less important. However, retrospective checking by skilled inspectors was required to detect faults.

3. The implications of the Hawthorne experiments caused a radical change in thinking about the treatment of workers. In particular, the use of Taylorism was criticised as being inefficient.

4. The period since the Second World War has been significant for two developments:

- Standards were used to control quality in the armaments industry – the forerunners of ISO 9000 (formerly BS 5750).

- Japanese industry rebuilt itself to become world-beating using methods which have become known as TQM, now imitated by Western industry.

QUESTIONS

Discussion

1. How difficult do you think the average construction organisation finds the concept of implementing Japanese management?

2. What would help the construction industry to learn the lessons of Rover's collaboration with Honda?

Individual

Write an essay of about 2000 words covering all three items:

(a) Contrast the advantages and disadvantages of mass production compared to more contemporary methods of managing worker relations.

(b) Why do some organisations continue to use Taylorism (fast-food restaurants are a notable example)?

(c) For how much longer do you think that Western producers would have continued to 'manage' in their traditional way if so many consumers had not switched to Japanese products?

FURTHER READING

Murta, K. and A. Harrison (1995) *How to make Japanese management methods work in the West*. Gower, Aldershot.

Oliver, N. and B. Wilkinson (1992) *The Japanization of British industry: new developments in the 1990s*. Blackwell, Oxford.

Sheldrake, J. (1996) *Management theory: from Taylorism to Japanization*. International Thomson Business Press, London.

Whitehill, A. M. (1992) *Japanese management: tradition and transition*. Routledge, London.

CHAPTER 4

THE GURUS OF QUALITY MANAGEMENT

OBJECTIVES

- Define management guru.
- Understand what gurus bring to quality management.
- Evaluate the impact of Deming and Juran on postwar Japan.
- Trace the development of more contemporary theories.

4.1 CONTEXT

Certain names will occur very often throughout this book. This is because in the world of quality management, they have become what are known as gurus. They are respected as being those who are able to offer special advice in implementing quality management. Such advice they believe will assist managers in organisations to achieve radical improvement. It is useful to know who the gurus are, what it is they offer, and why some appear to have achieved such an exalted status.

Contributors to the theory of quality management fall into three categories:

1. *Pioneers* are accepted as having been at the forefront of the quality revolution:

   - Deming
   - Juran

2. *Followers* were usually influenced to a greater or lesser extent by Deming and Juran:

   - Crosby
   - Imai
   - Feigenbaum
   - Ishikawa
   - Ohno
   - Shingo

3. *Contemporaries* came to prominence during the 1980s, following concern at the 'sudden' success of Japanese competition. They provided methods and philosophies

to 'cope' with the perceived threat of Japan. Most stressed the need for Western industry to become acquainted with the 'new' theories of quality which the Japanese had used to such great effect. They advocated what is generically called TQM. Some of the contemporaries are

- Peters and Waterman
- Pascale and Athos
- Ouchi
- Conway
- Schonberger

4.2   WHAT IS A GURU?

In most areas of management there will usually be a handful of individuals who have achieved significance by developing their supposition of how things should be done. Their ideas, or theories provide an indication to managers in organisations of what actions should be pursued in order to get 'desired' results. As Kramer explains, a management idea is

> derived from inductive and deductive reasoning. It is systematically organized knowledge applicable to a relatively wide area of circumstances. As a system of assumptions, accepted principles and rules of procedures . . . [it] assists managers to analyze and explain the underlying causes of a given business situation and predict the outcome of alternate courses of action. (1975:47)

Some argue that quality management is the latest trend in management thinking (Huczynski 1996:43), but quality management is not exceptional in having individuals offering theories on how best to manage organisations. However, in quality, gurus are significant in the fact that some have achieved superstar status. This is due to the belief that the 'cures' they offer hold out the possibility of creating radical organisational improvement, hence competitive market advantage.

If you look in a dictionary you will see the word *guru* translates as a spiritual teacher, often associated with the Hindu or Sikh faiths. What makes a guru important in both religion and management is that their word is to be obeyed. Their acquired knowledge makes the advice they offer very special. As Huczynski states: 'Guru theory took off at a time when managers appeared to need extra guidance and ideas. The rise of modern management guru theory can be dated to the early 1980s' (1996:42).

He goes on to explain how these gurus were often able to sell their ideas in books, and other media to the extent that their income matched that of film, television and pop stars (ibid.:2). This may help to explain the apparent surfeit of those who want to help managers.

However, at the beginning of the 1980s, there was a very good reason why managers of organisations in the United States and Europe needed the advice that gurus could provide. Certain events occurred in the 1970s which suddenly challenged the very markets upon which the organisations depended for survival. Among them were the oil price rises imposed by the Organisation of Petroleum Exporting Countries (OPEC). As a result, Western goods became more expensive when compared with international competitors. Coinciding with this was the influx of goods from Japan, goods which were cheaper and had better performance.

4.3 THE INFLUENCE OF JAPAN

The effect of Japan has been crucial to the development of quality management in the West (see previous chapters). Some began to believe that the only way to compete against the Japanese was to learn what they did and to adopt similar management approaches (Ouchi 1981; Pascale and Athos 1982). The need to learn what the Japanese had achieved made the message of the early quality gurus influential. This is particularly so in the case of Deming and Juran. It was the discovery of Deming in 1980 that was to lead to the interest in what is now called TQM (Micklethwait and Wooldridge 1996:7).

It is somewhat ironical that Deming and Juran were sent to assist the Japanese in their postwar rebuilding of industry. Crainer makes precisely this point: 'The failure of American corporations to listen to Deming and Juran has often been commented on. In retrospect it appears to be one of the century's most profound errors' (1996:141).

4.4 THE PIONEERS

4.4.1 Dr William Edwards Deming

Dr Deming (1900–1993) holds a special place in the history of quality management. This is because of his work in Japan where, since the 1950s, a prize named after him is awarded to organisations or individuals for excellence in the use of quality management. He is revered by some, who believe him to be the father of quality management: 'Deming has been called the founder of the third wave of the industrial revolution. His name has become synonymous with the reason for Japanese industrial success in the second half of the twentieth century' (Logothetis 1992:14).

Deming originally studied mathematical physics and obtained a PhD at Yale in 1928. During the summers he worked at the Hawthorne plant in Chicago, belonging to Western Electric. Hawthorne is the plant where many experiments were carried out, some of which led to the human relations approach. Walton makes the point in her description of the life of Deming that 'some of his ideas about management are rooted in his experience at Hawthorne' (1989:6).

Deming became an advocate of the use of statistical process control (SPC) for problem solving and quality improvement. However, it was Dr Walter Shewhart, a Hawthorne statistician, who originally developed the principles of statistical quality control (SQC). This provided the basis for Deming's work on SPC.

Shewhart's work was based on the idea that every task has variation, and that by using simple statistics to understand its nature, it is possible to try reducing it, perhaps to eliminate it altogether. As Shewhart believed there are two types of variation, uncontrolled and controlled; Deming called them special and common causes, respectively. Uncontrolled variation is unusual and does not occur randomly. It is often due to special circumstances which the person operating the equipment will be best able to recognise and solve. Controlled variation is inherent in the system. This type of variation is random; its occurrence and resolution are beyond the power of the operator. The system for production will need to be changed, and this must usually be decided by management.

The main significance of SQC is that, by being able to differentiate between the causes of variation, it is possible to know who is responsible for solving any problems that occur in a process. As Shewhart originally believed, only about 15% of variation is special, hence able to be solved by operators. The rest is common and is up to management to solve. 'Thus blaming workers is de-emphasized. . . . Management's task is to provide or develop the organizational system that enables workers to perform to maximum capacity' (Bell, McBride and Wilson 1994:94).

Any organisational system should also include a working environment where workers feel comfortable. The importance of this had been demonstrated by the Hawthorne experiments (Chapter 3). Shewhart's work was also sympathetic to these findings.

In 1931 Shewhart published his findings in a book called *Economic control of quality of manufactured product*. However, Deming, believed that SPC, as he called it, was capable of being applied to any process, regardless of whether or not it was manufacturing. Deming's perceived expertise as a statistician in the Department of Agriculture was soon recognised, and he was put in charge of the 1940 census, where the principles of SPC were applied. As a result of training and management in SPC, productivity in processes such as coding and card punching increased by up to six times. America, which had entered the war, benefited from Deming's assistance in the management of the war effort. Large increases in production, and reduction of waste and scrap were made possible through SPC.

Following the Allied victory, America's industry got back to normal. Producers were faced with demand for consumer goods which had not been available during the war. As a consequence, the preoccupation was on producing enough to satisfy this demand. SPC was forgotten, and scientific management was used instead. With its emphasis on deskilling, it certainly produced the desired output, but quality came to mean end-of-line inspection. If problems were found, there would be rework, incurring extra costs. This was acceptable because there was hardly any competition, and there was enough profit to cover any rework. Logothetis makes the point that, because Western managers believed this situation would last forever, they ignored what Deming had demonstrated by using SPC. Logothetis also believes that managers simply did not accept they were responsible for 85% of the problems.

Deming might easily have been forgotten, his contribution to the war effort meriting a footnote at most. But his crucial role in rebuilding Japanese industry meant that he became much more famous. After defeat in the war, Japan was under

American control. General Douglas MacArthur, Supreme Commander of the Allied Powers in Japan, recruited Deming in 1947 to assist in preparing a census in 1951.

Japan was a country in ruins. Industrialists were searching for a new beginning A group called the Union of Japanese Scientists and Engineers (JUSE) was formed to create the conditions for a revival in Japanese industry. They had been given copies of Shewhart's book by some Bell Telephone employees, who were in Japan to give advice. As a result of reading the book, they became interested in the technique of SPC.

Fortuitously, some members of JUSE had met Deming, and in March 1950 its managing director, Kenichi Koyanagi, wrote to Deming requesting him to deliver a lecture course about SPC to researchers, plant managers and engineers. Deming, who had become the metaphorical 'prophet without honour in his own country', was only too happy, and replied: 'As for remuneration, I shall not desire any. It will be only a great pleasure to assist you' (Walton 1989:12).

On 19 June 1950 he delivered the first lecture to a standing-room-only crowd of over five hundred. This was to continue wherever he spoke. It was common for people to be turned away. Because of his experience in America, where he had not been able to convince senior managers, Deming insisted on meeting with the *Kei-dan-ren*. This was the association of Japan's chief executives. He told his audiences, which consisted of the leaders of companies such as Sony, Nissan, Toyota and Mitsubishi, that all managers should use SPC. In conjunction with well-trained and well-educated workers, Deming proposed that SPC was essential to meet forthcoming demands by customers for high quality goods. As he stressed, 'The consumer is the most important part of the production line' (ibid.:14).

He also told them that, if they followed his advice, Japan would be so competitive in five years that manufacturers from the rest of the world would be 'screaming for protection'. As Deming later reflected with a certain sense of satisfaction, he was wrong; it took only four.

Walton describes how some of the Japanese found it hard to visualise Deming's approach: 'Some of those men would later tell years later that they thought his optimism was crazy, [but] at the time they had been willing to swallow their disbelief. In a sense, having lost all, they had nothing to lose' (ibid.:14).

The rest, of course, is history. Japanese manufacturers went from strength to strength. As well as naming a national quality prize after him, in 1951, Deming was awarded the Second Order of the Sacred Treasure by Emperor Hirohito in 1960. However, despite achieving such high honour, Deming remained unrecognised in America. This was until 1980, when an NBC documentary researcher, Clare Crawford-Mason, discovered him almost by chance. She had been working on a programme to investigate why America was 'suddenly' experiencing such an economic threat from Japan.

The programme *If Japan can, why can't we?* was broadcast on 24 June 1980. It included fifteen minutes dedicated to Deming, and described what he had done for the Japanese thirty years previously. It also described how Deming had been hired by William Conway, president of Nashua, a photocopier company based in New Hampshire. Because of Deming's advice, Nashua had saved millions of dollars and substantially increased productivity.

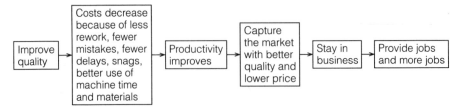

Fig. 4.1   The Deming chain reaction (Reprinted from *Out of the Crisis* (1986) by
W. Edwards Deming by permission of MIT and The W. Edwards Deming Institute.
Published by MIT, Center for Advanced Educational Services, Cambridge, MA 02139, USA.
Copyright © by The W. Edward Deming Institute)

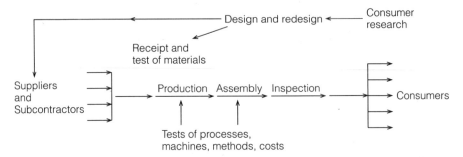

Fig. 4.2   The Deming flow diagram (Source, as for Fig. 4.1)

If Deming had been a prophet without honour, the television programme
changed all that. Deming was in demand by major companies. Not surprisingly,
those already experiencing the force of Japanese competition were first to require
his services, i.e. car producers like Ford and General Motors. They wanted him to
advise on how they could achieve what the Japanese had accomplished. So at the age
of 70, Deming returned to the work he had begun during the war. He published
Out of the crisis in 1986; it became a bestseller and is credited with starting the
quality revolution that caused major organisations in the West to re-examine their
methods of operation. He continued to preach his message up to his death in 1993,
and despite the improvements many Western organisations were making, he still
believed that not enough was being done. Deming's beliefs are summarised in
Figures 4.1 and 4.2.

Deming also believed in the need to have a method for ensuring that quality is
incorporated into every task an organisation carries out (no matter how trivial).
He provided what is now known as the Deming cycle (although he called it the
Shewhart cycle). It consists of four vital steps that should be applied continuously
and repetitively to every task. As Logothetis states, this cycle provides the basis for
'problem-solving and loop-learning' (1992:55); it is shown in Figure 4.3.

The various stages are obvious. Any activity should be planned before
implementation. There will need to be evaluation using available data to select the
best way of implementing any task. The doing will require every person involved to
implement the plan, but also to collect data on what occurs during this stage. The

Fig. 4.3 The Deming plan, do, check, action (PDCA) cycle (Reprinted with the permission of Blackwell Publishers from Dale and Cooper 1992)

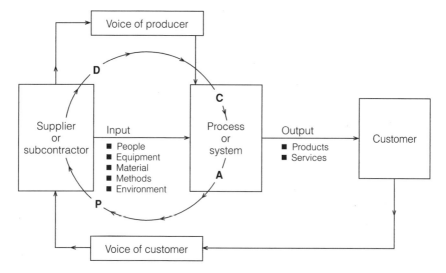

Fig. 4.4 The PDCA cycle and a system for ensuring customer satisfaction (Reprinted with the permission of Prentice Hall Europe, from Logothetis 1992:56)

collection of data allows checking to be carried out. It is then possible to evaluate how effective the plan was. If mistakes occurred or deviation was necessary, the reasons and objective data will assist in understanding why. The final part, act, is where corrective action, following the learning that has occurred, can be incorporated into future planning of tasks. The aim is to ensure that past mistakes are not repeated. The emphasis is to use the cycle continuously. Improvement is a never-ending objective.

Figure 4.4 shows how the Deming cycle (PDCA) can be used in a system which explicitly comprises all the parties up to and including the customer. The aim will

Table 4.1 Deming's fourteen points

---

1. Create constancy of purpose towards improvement of product and service, with the aim to become competitive, stay in business and provide jobs.

2. Adopt the new philosophy – we are in a new economic age. Western management must awaken to the challenge, learn their responsibilities and take on leadership for future change.

3. Cease dependence on inspection to achieve quality. Eliminate the need for inspection on a mass basis by building quality into the product in the first place.

4. End the practice of awarding business on the basis of price tag. Instead, minimize total cost. Move towards a single supplier for any one item on a long-term relationship of loyalty and trust.

5. Improve constantly and forever the system of production and service, to improve quality and productivity, and thus constantly decrease costs.

6. Institute training on the job.

7. Institute leadership (see point 12): the aim of supervision should be to help people, machines and gadgets to do a better job. Supervision of management, as well as supervision of production workers, is in need of overhaul.

8. Drive out fear, so that everyone may work effectively for the company.

9. Break down barriers between departments. People in research, design, sales and production must work as a team, to foresee problems of production and problems in use that may be encountered with the product or service.

10. Eliminate slogans, exhortations and targets for the workforce which ask for zero defects and new levels of productivity. Such exhortations only create adversarial relationships, as the bulk of the causes of low quality and low productivity belong to the system and thus lie beyond the power of the workforce.

11a. Eliminate work standards (quotas) on the factory floor; substitute leadership instead.

11b. Eliminate management by objectives, by numbers and by numerical goals; substitute leadership instead.

12a. Remove barriers that rob the hourly worker of his or her right to pride of workmanship. The responsibility of supervisors must be changed from sheer numbers to quality.

12b. Remove barriers that rob people in management and in engineering of their right to pride of workmanship. This means, *inter alia*, abolishing the annual or merit rating and management by objectives.

13. Institute a vigorous programme of education and self-improvement.

14. Put everybody in the company to work to accomplish the transformation. The transformation is everybody's job.

Source: Reprinted with the permission of Prentice Hall Europe, from Dale, Boaden and Lascelles (1994:16).

be to ensure that customers get what they want. Deming also provided a list of fourteen points which must be adopted by any organisation wishing to follow his lessons. They are shown in Table 4.1.

Many point to the fact that, despite Deming's use of seemingly technical tools such as SPC, his fourteen points provide a basis for radical change, often called cultural change. This shift, which requires people to radically alter their personal thinking and attitude, is at the heart of TQM.

4.4.2 Dr Joseph Juran

The life of Joseph Juran has mirrored that of Deming. However, as Crainer states, he has 'lived somewhat in Deming's shadow' (1996:237). Juran was born in the Balkans, but emigrated to America, where he worked as an inspector of quality for the Western Electric Company. Like Deming, he was similarly influenced by the work of Shewhart. Juran believed there had to be an alternative to the accepted methods which would avoid producing large amounts of scrap. His answer was to analyse the process of production and to avoid problems before they occurred. His ideas were also ignored, just like Deming's, and he too found recognition in Japan.

Three years after the publication of his book, *Quality control handbook* (1950), Juran accepted an invitation from JUSE to lecture in Japan. However, his message was different from Deming's. Deming provided a philosophy whereas Juran advanced a method which was less statistically driven. He developed the concept of company-wide quality management (CWQM). This was a means of disseminating quality through the whole organisation. As he stressed, every person in an organisation is responsible. This, it can be argued, was an early reference to the concept of empowerment (Chapter 8).

Another major difference in Juran's beliefs, compared to Deming, is his conviction of the need to reduce the cost of quality. 'The language of top management is money' (Dale, Boaden and Lascelles 1994:19). So quality, which he defined as fitness for use, requires quality of design, quality of conformance, availability and adequate service. It is therefore incumbent on the senior managers of any organisation to strive towards increased performance and reduced costs. Juran provides ten essential stages to achieve this objective:

1. Create awareness of the need and opportunity for quality improvement.
2. Set goals for continuous improvement.
3. Build an organisation to achieve goals by establishing a quality council, identifying problems, selecting a project, appointing teams and choosing facilitators.
4. Give everyone training.
5. Carry out projects to solve problems.
6. Report progress.
7. Show recognition.
8. Communicate results.
9. Keep a record of successes.
10. Incorporate annual improvements into the company's regular systems and processes, and thereby maintain momentum.

Although Juran lectured to senior managers in Japan, the workers were also eager to find out what he said. His lectures were translated and sold at newspaper kiosks and they were broadcast on national radio. Many large organisations had started programmes

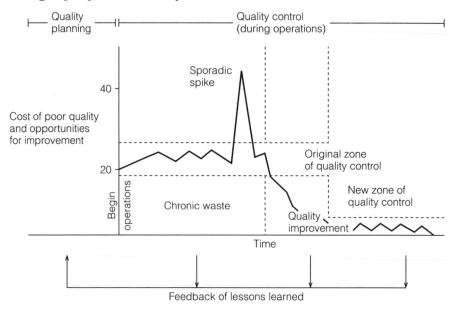

Fig. 4.5 The Juran trilogy (Reprinted with the permission of Prentice Hall Europe, from Logothetis 1992:63)

of literacy and Juran's lectures were included. Literacy sessions formed the basis for quality circles, a very important aspect of the Japanese approach to quality management: 'It was a short step from the "reading circles" becoming problem-solving quality improvement meetings . . . quality circles were created' (Bank 1992:71).

Juran suffered the same fate as Deming, he was ignored in America. Similarly, it took the impending problems of the late 1970s before America eventually recognised him in the early 1980s. Indeed, like Deming, Juran has spent his retirement preaching to those who are now very willing to listen.

Deming had fourteen points, Juran offers three. The Juran trilogy links together planning, control and improvement (Figure 4.5).

Planning

Juran firmly believes that quality does not happen by chance; it needs to be planned. This will require that certain measures are instituted, especially training. As he recommends, every member of the organisation must be able to contribute to the planning effort. Juran believes in carrying out the following steps:

1. Identify the requirements of all the customers in the process. This obviously includes the end recipient (external) but also intermediate customers (internal).
2. Ensure that customers' needs, as identified above, are translated into language that every person can easily understand.
3. Ensure the process will produce something that meets these needs.
4. Produce the product for the customer.

Control

This part of Juran's trilogy is in line with Deming's belief that, once it is in operation, a process must be continuously monitored for variation. Any process should be under control. Juran also agrees with Deming in that management must take responsibility for the majority, i.e. at least 80%, of problems that occur during production.

Improvement

Juran believes the managers in any organisation have a responsibility to ensure that all necessary steps are taken to improve the system in which every person works. He also stresses that much will depend on getting people to recognise the importance of changing their attitude. This he believes requires those at the top to create an environment where cultural change occurs.

Although Deming and Juran have much in common, they themselves argued that their approaches were very different. Deming concentrates on education, whereas Juran concentrates on implementation. Logothetis (1992) believes the explanation is that Deming was mainly involved with academics and consultants, Juran with professional management.

Juran is still active at the time of writing. He remains dismissive about attempts by those in the West to deal with quality management. According to him, quality is often not given enough importance by senior managers. As a result, their efforts to match Asian competitors will be limited.

4.5 THE FOLLOWERS

4.5.1 Philip Crosby

Crosby is another American quality guru who came to prominence because of his book, *Quality is free*. It was published in 1979, just before the rediscovery of Dr Deming in the West.

Crosby's book sold over a million copies, allowing him to leave his job as vice president of ITT and set up his own consultancy. Although he was at ITT, he had special responsibility for quality. It is reputed that his programme for quality saved $720 million in one year using techniques Crosby instituted.

Crosby's central message was that by 'doing it right first time', like the Japanese, it was possible to build in quality. This meant quality could be free, but

> quality is not only free, it is an honest-to-everything profit maker. Every penny you don't spend on doing things wrong, over or instead becomes half a penny right in the bottom line. In these days of 'who knows what is going to happen to our business tomorrow' there aren't many ways let to make a profit improvement. If you concentrate on making quality certain, you can probably increase your profit by an amount equal to 5 to 10 per cent of your sales. That is a lot of money for free. (1979:1)

Some people think Crosby had as much influence as Deming and Juran. Kennedy (1994) believes the simplicity of Crosby's message makes it very persuasive. Bank reinforces this belief by stating that Juran's zero defects goal, one of his so-called absolutes, is 'practical, reasonable and achievable' (1992:76).

Crosby uses medical metaphors to teach his message; an organisation that does not produce quality output is like a 'patient in need of a vaccine.' Customer dissatisfaction is the most obvious, and final symptom of a diseased organisation. He has identified five other symptoms of a 'sick' organisation:

1. The need to have an extensive system of field service to fix things that have gone wrong once the customer starts using a product or service.
2. What the customer gets, deviates from what they asked for.
3. Lack of management commitment to providing an environment where quality is positively encouraged.
4. Management deny any problems are theirs.
5. Management fail to measure the cost of non–conformance, or they refuse to accept the figures.

As Crosby suggests, senior managers of diseased organisations will often do four things:

1. Concentrate any effort to improve at the lowers levels, as though quality is nothing to do with them.
2. Call the improvement effort a programme rather than a process. A programme, Crosby stresses, indicates something temporary.
3. Blame others when things go wrong. Crosby believes in accepting that some experimentation will be required along the way, and that sometimes mistakes will be made.
4. Expect that improvement is easy and quick.

Crosby's 'vaccination serum' is a three-part remedy:

1. Uphold integrity and be dedicated to customer satisfaction.
2. Design any systems and operations so they improve quality.
3. Make communication paramount to encourage trust and cooperation between all parties involved, including suppliers and subcontractors.

Crosby also prescribes the proper management attitude for administering his serum:

1. Determination to succeed.
2. Preparedness to become educators or facilitators rather than the traditional 'controllers of labour'.
3. Willingness to create an atmosphere where workers are motivated to achieve their best efforts in a system which supports them.

Crosby's approach is summarised in his four absolutes for quality management:

1. Quality is conformance to requirements, not 'goodness' or 'elegance'.
2. It is always cheaper to do a job right first time.
3. The measurement of quality is the price of non–conformance.
4. The quality performance standard is zero defects.

Crosby also provides a list of fourteen points which encapsulate all of the above. Like Deming's fourteen points, he believes his list is mandatory for any organisation wishing to improve.

1. Demonstrate management commitment in every possible way.
2. Encourage every employee by using quality improvement teams.
3. Use quality measurement to show what needs to be done, and to indicate progress.
4. Evaluate the cost of quality to demonstrate any savings when they start to occur.
5. Use quality awareness to remind people what they are expected to contribute.
6. Aim to prevent problems by taking corrective action. This should be a value-adding exercise. Corrective action which is only about rectification, a standard approach for many organisations, will always be cost-adding. That is why Crosby believes retrospective correction is wrong.
7. Establish a committee for the zero defects initiative.
8. Educate employees and give them appropriate training.
9. Arrange a special day for discussion and presentation of awards – zero defects day.
10. Set goals to provide targets for improvement.
11. Remove the causes of error by making improvements and communicating them to all concerned.
12. Recognise the efforts of those who make the greatest contributions.
13. Formulate future policy in a quality council consisting of representatives from all areas of the organisation.
14. Do it all over again.

4.5.2   Masaaki Imai

Imai is a Japanese quality consultant who wrote a book in 1986 which drew attention to the word *Kaizen*. The word *Kaizen* is Japanese, and although it appears to have no precise definition, it is accepted to mean continuous, small-step improvement. It is the central tenet of TQM.

Imai's book, which undoubtedly attracted attention because of its title, *Kaizen: the key to Japan's competitive success*, is regarded as being the most superior publication explaining what Japanese management actually is (Bank 1992:191). Imai brought together all the management philosophies, theories and techniques that the Japanese have employed since the Second World War. Imai believes:

> The message of KAIZEN is that not a day should go by without some kind of improvement being made somewhere in the company. . . . The belief that there should be unending improvement is deeply ingrained in the Japanese mentality. (1986:4–5)

Imai also advocates the use of total quality control (TQC), also called company-wide quality control (CWQC). He admits that the concept of statistical quality control as a means to achieve improvement was brought to Japan by Deming and Juran.

Another view, perhaps somewhat revisionist, suggest the Japanese 'miracle' would have happened without Deming and Juran. However, what the Japanese did was to fully embrace, develop and improve their theories, something the West was not prepared to do at that time.

Kaizen is crucial to improvement because, according to Logothetis, it is 'a people-orientated approach which promotes discipline, participation, and involvement, skill development, morale and communication' (1992:91). He also stresses how it seeks to ensure that past improvements are maintained and future improvements are attempted. This is the essential difference between Japan and the West. Sid Joynson, a passionate advocate for people-orientated solutions, argues, 'In the West we have a "big leap at a time" mentality, a staircase concept of progress' (Joynson and Forrester 1995:145).

Joynson goes on to describe how the Western system tends towards large steps, usually accompanied by much effort. Once the step is complete, everyone relaxes and any potential improvement is reduced: 'Contrast this with the Japanese approach. Using Kaizen techniques put in the hands of workforce teams [the aim] is for continuous improvement. One small step after another, week after week' (ibid.:145).

4.5.3   Armand V. Feigenbaum

Feigenbaum wrote about TQC in a 1951 book entitled *Quality control*. He revised it ten years later, and added the word *total*.

Feigenbaum was formerly responsible for manufacturing at General Electric; now a quality consultant, he too is associated with the concept of *Kaizen* and has long advocated a total approach to quality which, he stresses, must involve every person. Feigenbaum believes quality should be built in, prevented, rather than inspected out. He thinks quality is nothing more than the best way of managing any organisation in a way that is

- Customer orientated
- Cost-effective
- Involves all the people who are committed to achieving the best

He goes on to provide a four-step approach:

1. Set quality standards.
2. Appraise the performance of quality standards.
3. Act when quality standards are not met.
4. Plan for improvement in quality standards.

He also stressed that management must commit themselves to three objectives:

1. Strengthening the quality improvement process at every opportunity.
2. Making sure that improvement becomes a habit (part of the 'new' culture).
3. Recognising that quality and cost saving are complementary objectives.

He identified three major types of cost (Feigenbaum 1956):

- Appraisal
- Prevention
- Failure

Perhaps constituting 25–30% of sales or operating costs, according to Feigenbaum, these three categories could be substantially reduced by TQC.

4.5.4   Kaoru Ishikawa

Ishikawa, who died in 1989 at the age of 74, is regarded by many as the person who developed quality circles. Quality circles are a way of bringing together all those who directly contribute to any tasks. As Ishikawa believed, the most important thing is that problems should be identified and solved by those directly involved. Like Deming and Juran, by whom he was influenced, Ishikawa believed that management bore most responsibility for improving the system in which everyone operated. In order to do this, senior managers need to understand what the problems are. Being involved in the group that identifies and solves the problems can only help managers to appreciate the importance of their role in managing the system better.

A quality circle will typically consist of a number of people who come together on an informal basis to problem solve. Because the group is not a device of management, it consists of representatives from those who carry out operations. It also benefits from contributors that are not directly involved in production; administrators are frequently included. It is also highly beneficial to involve suppliers and subcontractors.

The main objective of quality circles is to allow any problem solving to occur in an atmosphere of openness and cooperation. Accordingly, even though senior managers may attend, they are equal in status to everyone else. If a leader or spokesperson is required, they will be elected by the group. However, any managers who do attend, and are therefore fully aware of what has been achieved, will probably feel committed to the solution. This is useful if the solution needs to be communicated to other departments.

Ishikawa was an advocate of company-wide quality, where every person should be involved. In order to help people with problem solving, Ishikawa developed certain tools. These tools, which he stressed must be used by people who have been trained and educated to understand the importance of quality, were explained in his book, *What is total quality control? The Japanese way* (translated in 1985). His seven 'indispensable' tools are

- Fishbone diagrams
- Pareto analysis
- Stratification
- Tally charts
- Histograms
- Scatter diagrams
- Control charts

The use of quality tools to problem solve will be explained in detail in Chapter 10. Like many of those who advocate statistical techniques, Ishikawa believed it is important to know the magnitude of any problem, so its consequences must be measured using the seven quality tools. The effectiveness of any solution can also be investigated by continuous monitoring.

4.5.5   Taiichi Ohno

Ohno, who worked at Toyota Motor Corporation, pioneered what is widely known as just-in time (JIT). JIT is also called *Kanban*, which means a signpost or label. Signposts and labels are attached to items as they are sent through the stages of production. When all the items are used and more are required, the *Kanban* is used to instruct an operator to send some more. This can apply within an organisation or to external suppliers and subcontractors.

The main principle of JIT is that the operator will only act when a *Kanban* is received. *Kanbans* will thus only be sent as and when the item is needed for immediate use in the next production stage. The aim is to eliminate the seven *mudas* (non-value-adding aspects):

- Excess production
- Waiting
- Conveyance
- Motion
- The process itself
- Inventory
- Defects

The aim, therefore, of both workers and management is to concentrate collective effort on producing 'right first time'. This is another of the tenets of quality management. Together, workers and management identify all the activities which do not add value, hence are likely to inhibit continuous production. As Ohno stressed, for JIT to work, everyone involved must be trained to understand the need to get 'it right first time'. It is also essential, he believed, that the proper tools and equipment are used. Ohno is also credited with being the driving force behind the principles of lean production (Chapter 12).

4.5.6   Shigeo Shingo

Shingo has become associated with *Poka-yoke*. A concept he pioneered in Japan during the 1960s, it was born form a personal concern that the use of SPC was retrospective. Shingo believed in identifying errors as they occur, correcting them there and then.

Poka-yoke, also called defect zero, is closely related to *Kanban*. As Flood suggests (1993:28–29), because humans are 'inconsistent' and sometimes make errors, monitoring will best be achieved by automation. As a result, when the system using this technique finds an error, it will alert those concerned with solving it. Humans are therefore relieved of monotonous checking. Instead, they can concentrate on improvement and innovation. Consequently, over a period of time, the system should become almost error-free and any error which does occur will then be exceptional.

4.6 THE CONTEMPORARIES

4.6.1 Tom Peters and Robert Waterman

Tom Peters is regarded as the world's leading management guru, 'a star platform performer, commanding international fees to match' (Kennedy 1994:11). However, it was a coauthored book that brought him to such celebrity, and reputedly allows him to charge $65 000 per day for a seminar in America ($95 000 if he is requested to go abroad).

The book which Peters wrote with Robert Waterman in 1982 was *In search of excellence: lessons from America's best-run companies*. As Wilkinson and Willmott suggest, it was the interest that had been aroused in the early 1980s, into the 'secrets' of Japanese success, which created the market for a book that 'sought to identify ways of responding' to the challenge so lucrative:

> While the contribution of a strong corporate culture was widely taken as Peters and Waterman's chief lesson, it is relevant to note that they also stressed the importance of quality, in a way that anticipated the writings of more recent champions of quality management. (1995:5)

Micklethwait and Wooldridge note how *In search of excellence* has helped to stimulate a 'guru industry' which is now worth at least $750 million per year in America (1996:8).

Peters and Waterman worked as consultants for McKinsey. Together with Richard Pascale and Anthony Athos (see below) they carried out research into sixty-two of America's most successful companies. As a result of their research, Peters and Waterman came to believe that organisations require eight key attributes in order to become excellent:

1 *A bias for action* means the ability to respond quickly to new opportunities.
2 *Close to the customer* is the desire to provide superior quality and service, based on being intimately aware of what customers require.
3 *Autonomy and enterprise* requires an open communication system and an environment that encourages freedom of individuals to please the customer is encouraged. Customer-pleasers are called champions.
4 *Productivity through people* means people are regarded as partners, not merely as employees.
5 *Hands-on* managers get out of their offices and see what the workers do. *Value-driven* organisations have a set of values that should be supported by every employee.
6 *Stick to the knitting* and avoid trying to do things you are not good at.

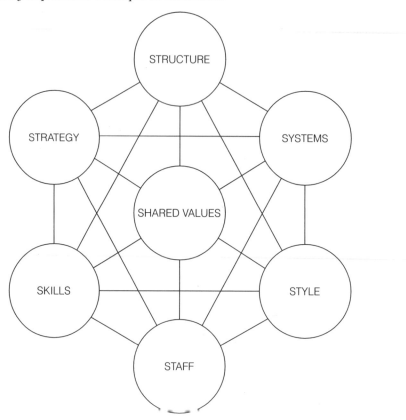

Fig. 4.6 The McKinsey 7-S framework (from *In Search of Excellence* by Thomas J. Peters and Robert H. Waterman, Jr. (1982:10) Copyright © 1982 by Thomas J. Peters and Robert H. Waterman, Jr. Reprinted by permission of HarperCollins Publishers, Inc.)

7 *Simple form, lean staff* means the organisation has the smallest possible structure and all its staff are doers; a complex organisation will usually require too many administrators.

8 *Simultaneous loose–tight properties* allow an organisation the freedom to act in its best interest but within the market's constraints.

In search of excellence was written using language untypical of previous management texts. This was a deliberate choice by the authors. Peters, in particular, is regarded as an evangelist. He preaches to managers in a way that the unconverted may find strange. As Crainer points out, his language seeks to exhort by being 'exciting' (1996:112).

Peters and Waterman also provided the McKinsey 7-S Framework (Figure 4.6).

According to Crainer, these seven S's provide 'a kind of instant guide to issues and topics which occupy managerial minds' (1996:113). They come in two sorts. The 'harder' elements are strategy, structure and systems. These are the 'technical' aspects of an organisation, and are essential to provide the right environment to support the other four. The remainder, staff, style, shared values and skills are 'soft'; they concern people, Peters and Waterman stress they are essential for an excellent organisation.

Here are all seven:

1. *Strategy* ensures the resources are dedicated to achieving goals that create superior quality.
2. *Structure* must be appropriate to support the efforts of those who work in the organisation.
3. *Systems* should enable rather than disable.
4. *Staff* require education and training to achieve goals.
5. *Style* should be used by managers to create the right culture.
6. *Shared values* mean that all employees (partners) agree the aims and objectives to be supported.
7. *Skills* are acquired by all employees through appropriate education and training.

After publication Peters and Waterman were able to set up their own consultancies. Each continues to write, but it is Peters who has achieved the greater fame, or perhaps notoriety, as 'the most famous management guru in the world' (BBC 1993). As well as being able to deliver up to sixty of his seminars per year, Peters is also a prolific writer on management. Waterman, who is less well known than Peters, continues to write on the subject of excellence.

4.6.2   Richard Pascale and Anthony Athos

Pascale and Athos, who worked with Peters and Waterman, are remembered for having published *The art of Japanese management* a year before *In search of excellence*. They provided the '7-S framework which also appeared in the Peters and Waterman book. However, they have not been as successful as Peters and Waterman.

According to Huczynski, the reason for Pascale and Athos's relative lack of success, compared to Peters and Waterman, may be because their message was regarded as too pessimistic. In particular, they drew attention to American companies' apparent inability to compete with Japanese competitors. They suggested that America's managerial decline was an effect of culture, and that Japanese culture was 'more attuned to coping with the inherent ambiguity, uncertainty and interdependence of organizational behaviour' (Thomas 1993:142). American culture, they argued, saw these indefinables as negative, so it tried to eliminate them. Besides, stressed Pascale and Athos, America encouraged independence rather than interdependence. As Thomas explains, they also suggested that Japanese organisations were more successful than their American competitors, because although both were 'equally adroit at the hard Ss, the Japanese were particularly skilled in the 'soft' ones' (ibid.:142).

4.6.3   William Ouchi

Another book which drew attention to Japanese management, and predated *In search of excellence*, is *Theory Z* by William Ouchi. Its ideas owe much to the social psychologist Douglas MacGregor, who is chiefly remembered for motivation theories X and Y. Theory Z, which aimed to coordinate the aspirations of the individual with the requirements of the corporation, was being formulated by MacGregor when he died. Ouchi took the unfinished work and developed it to explain seven major characteristics which, he believed, were contributing to America's industrial decline:

1. *Short-term employment* contributed to impersonal relationships and lack of loyalty.
2. *Rapid evaluation and promotion* were caused by short-term employment and they meant that new recruits had to be judged quickly. If they were good, the idea was to promote them fast in case they left.
3. *Specialised career paths* meant that people acquired a specific skill which they could take from organisation to organisation in search of higher pay.
4. *Explicit and formal control mechanisms* were formal procedurised systems used by organisations for the purpose of control.
5. *Individual decision making*, where individuals must take responsibility for decisions they must make in relation to others.
6. *Individual responsibility* was coupled with individual decision making to encourage the brave, or perhaps the reckless.
7. *Segmented concern for employees* concentrated working relationships on the workplace itself and did not encourage shared outside interests.

As Ouchi pointed out, the most successful Japanese organisations had characteristics that were the complete opposite. He called them Theory Z organisations, and put the emphasis on things like lifetime employment along with much more consensus and sharing of decision making and responsibility. Ouchi also argued that it was concern for people which contributed to intimacy, trust and understanding. These, he explained, would lead to everyone putting maximum effort into making their organisation excellent.

4.6.4   William Conway

With regard to Deming, Conway can say, 'I told you so!' As president and chief executive of the Nuasha Corporation, he had personally decided to invite Dr Deming to set up a total quality programme in 1979. In the 1980 broadcast which alerted America to Dr Deming, Conway reported how Deming's advice was saving Nuasha millions of dollars and substantially increasing productivity.

Having watched him on the television, Conway followed Deming's advice and set up his own consultancy. His approach is that the 'right way to manage' incorporates technology for continuous improvement. This requires the development, manufacture, administration and distribution of consistent low-cost products and services, which customers require and will be fully satisfied with.

Conway believes there are three areas of waste:

- Time
- Capital
- Material

Efforts at quality management should aim to reduce them. In order to do this, Conway provided the following advice:

1. Value the human element, i.e. people; train and encourage staff at every opportunity.
2. Use statistical process control (SPC) to find out what customers (internal and external) require and what employees require. And use statistics to assess the effectiveness of all equipment and systems. Always aim to improve in the future.
3. Make sure that everybody solves their problems using simple statistics.
4. Make sure that managers are imaginative in their thinking; use vision statements and brainstorming.
5. Encourage managers to try industrial engineering; use improvement techniques such as work study, work simplification, method analysis and material handling.

4.6.5 Richard Schonberger

In his 1990 book *Building a chain of customers*, Schonberger argues that quality has progressed 'up' four levels since the 1980s. These are corrective, preventive, cost-based and customer-orientated. Schonberger argues that the next step is for all employees to become 'owners' of efforts towards continuous improvement. He provides twelve multiple dimensions of quality which will become essential in the future:

- Conformance to specification every time
- Superior performance
- Quick response
- Quick-change expertise
- Innovative features
- Reliability
- Durability
- Serviceability
- Aesthetics
- Perception of quality
- Humanity
- Value

Customers that perceive that they receive quality may remain loyal regardless of price. Humane organisations acknowledge their customers see honesty and friendliness as part of the service. Although all twelve are extremely important, Schonberger believes that value is a vital issue, and world class organisations are continually striving to improve their value. Chapter 12 describes the concept *world class* in greater detail.

SUMMARY

1. So-called gurus of quality management are important because of their contribution to developing theories and because they advise organisations who wish to implement quality management.

2. The most influential gurus are the pioneers, Deming and Juran. This is because of their influence and assistance, which contributed to the rebuilding of postwar Japan. In particular, they stressed that quality must be based on the people who work in the system or organisation.

3. They had several followers, mainly Japanese, who continued the quality revolution in Japan. Crosby is the notable exception, and is remembered for his book, *Quality is free.*

4. Following the 'discovery' of Dr Deming, TQM was invented in response to the threat of Japanese imports, regarded as superior even though they cost less. TQM encompasses many approaches to quality improvement and seeks to learn from what the Japanese have achieved using quality management. Out of the need to learn TQM there have emerged several contemporary thinkers. All of them give advice and guidance on how to achieve radical organisational change, leading to improvement hence competitive advantage.

QUESTIONS

Discussion

1. Without the threat of Japanese imports, would the West have embraced TQM?

2. Has history taught us any lessons about listening to prophets in their own countries?

Individual

Write an essay of about 2000 words covering all three items:

(a) Compare and contrast the thinking of Dr Deming's philosophy with one of the more contemporary gurus of quality management. Draw some conclusions.

(b) Why do many organisations apparently find Deming's fourteen points so difficult to comprehend in terms of practical application?

FURTHER READING

Crainer, S. (1996) *Key management ideas: thinking that changed the management world.* Pitman, London.

Huczynski, A. A. (1996) *Management gurus: what makes them and how to become one.* International Thomson Business Press, London.

Kennedy, C. (1991) *Guide to the management gurus: shortcuts to the leading management thinkers.* Century, London.

Kennedy, C. (1994) *Managing with the gurus: top level advice on 20 management techniques.* Century, London.

Micklethwait, J. and A. Wooldridge (1996) *The witch doctors: what the management gurus are saying, why it matters and how to make sense of it.* Heinemann, London.

CHAPTER 5

THE USE OF QUALITY ASSURANCE

OBJECTIVES

- Describe what QA really means.
- Appreciate that ISO 9000 is the international standard for quality management.
- Explain the requirements of ISO 9000.
- Give a practical example of each clause in ISO 9000.

5.1 CONTEXT

Much has been written in recent years about the introduction of QA using quality systems such as ISO 9000 (formerly BS 5750). The result has been debate, usually conducted on the basis of either defence or condemnation of QA.

This chapter describes what QA is intended to do, and what it can do, if implemented and managed in a way that seeks to support the efforts of those who use it. However, using QA as a 'weapon' to ensure that employees 'do what they are told' is likely to result in it being resented. It will also require someone to 'police' the system.

5.2 WHAT IS ISO 9000?

Chapter 2 dealt with definitions of quality management. It assumed that all organisations aim to provide their customers with the 'best'. In practice this does not always happen. There may be good reasons, and customers are normally free to choose whether or not they accept them. If there are no good reasons and if alternative suppliers exist, then it will not be surprising if customers look elsewhere.

Any organisation will therefore be well advised to do all it can to ensure it satisfies its customers. Quality assurance is something that can assist in achieving this. BS EN ISO 8402 *Quality management and quality assurance – vocabulary* defines it as 'all the planned and systematic activities implemented within the quality system, and demonstrated as needed, to provide confidence that an entity will fulfil requirements for quality' (1995:25).

The choice of activities is entirely up to the organisation. If the organisation is small, or its processes extremely simple, its QA may be on accepted practice, i.e. do what has always been done. Usually by word of mouth, this was the way that skilled workers traditionally learned their trade.

In practice, and especially in construction, organisations can be relatively small, but 'act large'. This means the range of work can be varied, and sometimes extremely complex. The use of informal methods is usually inadequate. Then a quality system becomes a way of demonstrating that formal mechanisms exist for dealing with different situations and complexity.

Quality systems are defined in BS EN ISO 8402 as 'organizational structure, procedures, processes and resources for implementing quality management' (ibid.:26). Their operation is explained by Dale: 'The purpose of a quality system is to establish a framework of reference points to ensure that every time a process is performed, the same information, methods, skills and controls are used and applied in a consistent manner' (1994b:334).

An alternative definition comes from Hughes and Williams: 'saying what you do, doing what you say, [and] recording that you have done it' (1991:6). Accordingly, any organisation can have a quality system. Simply specify how something is to be done then verify whether it has been achieved.

Every task, no matter how simple, has certain steps which need to be carried out. Normally they will be accepted as the best way of doing it. And they may have been written down. This means a quality system can be applied to every operation. The result is normally a collection of procedures and work instructions. It is also highly probable that most organisations have them before implementing QA; they are called standard management procedures.

5.3   THIRD-PARTY ASSESSMENT

This is the crucial difference. Verification, or recording what you have done, is probably where QA departs from merely having standard procedures. There are several reasons why it is essential to demonstrate standard procedures have been adhered to.

Firstly, a customer may reasonably ask for proof. An organisation might try to argue that self-assessment (first-party assessment) provides the necessary proof. However, customer assessment (second-party assessment) may be possible. This will necessitate potential customers visiting the premises of suppliers and subcontractors. It used to be standard practice by some large clients, but it was expensive and extremely time-consuming, especially where many suppliers were involved.

A better way of providing proof of adherence to standard procedures is to employ an expert to carry out the assessment on your behalf – third–party assessment. The problem is that, where organisations adhere to their own systems, there can be great variation.

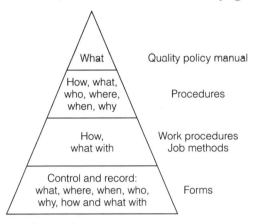

Fig. 5.1   Quality system documentation (Wilton 1994:47)

Clause 4.2: quality system

Set-up and maintenance of a quality system to meet customer requirements should include the following items:

1. *Documentation (clause 4.5)*: the implementation of procedures, instructions and all other aspects required to operate the system.
2. *The quality manual*: this brings together the standard elements, such as procedures (see below), and describes how they operate in order to maintain organisational effectiveness. It provides overall guidance on consistent application of the quality system. Its language should be simple and its layout should be clear. Figure 5.1 illustrates some typical documentation. A procedure is something which tells someone how to carry out their task. The activities in a QA procedure are specified in a way that allows their assessment (auditing) once they have been completed.
3. *Planning*: this shows how the system will be implemented and consistently applied. It is crucial in any managerial activity and will be best covered by a general statement of intent. Quality plans may also be considered. A quality plan sets out the required quality practices, resources and sequence of activities to be completed for a particular product or service. Quality plans are normally used where the output of the organisation is frequently varied to suit individual customer needs (clause 4.3). As such, they tend to be widely used in construction in order to provide a way of tailoring the system to provide QA that ensures customer needs are fulfilled on individual projects. Where this occurs, it is sensible to refer to their use, and how they will operate in the quality manual.

Clause 4.3: contract review

Contract review is at the heart of QA; it meets the need to ensure customer satisfaction. The word *review* may be misleading, because it implies retrospection. This part of ISO 9000 is concerned with having a recognised and accepted method (possibly incorporated into a procedure) to ensure the requirements of the customer are adequately defined and understood, and to ensure the organisation can fulfil them.

The use of quality plans to suit individual clients is a sensible approach for construction (Wilton 1994:49). Their purpose is to clarify all the necessary issues. They may form part of the legal contract between the parties. There may also be arrangements for testing and inspection on particular parts the work (clause 4.10).

QA requires no more than a demonstration that the process under review is actually taking place. It would be surprising if an organisation had not already considered how they can best satisfy their customers.

Clause 4.4: design control

Design control occurs only in ISO 9001. It is the longest clause and is often regarded as the most daunting. It recognises that design, if carried out, is very important. Consequently, it deals with all the essential issues. The first two elements are concerned with what to do before starting:

1. *Design and development planning (4.4.2)* ensure that adequate resources are available.
2. *Organisational and technical interfaces (4.4.3)* ensure the organisational aspects are appropriate, especially where different departments are involved.

The following elements deal with the actual process:

3. *Input (4.4.4)* ensures the customer specification is fully understood and all statutory requirements are incorporated.
4. *Output (4.4.5)* ensures that input can be expressed.

The remaining elements deal with the checking of design and changes:

5. *Review (4.4.6)* ensures the design actually meets the requirements as set in the input.
6. *Verification (4.4.7)* ensures the design will produce what was originally intended.
7. *Validation (4.4.8)* ensures the actual result of the design fulfils the customer's requirements.
8. *Changes (4.4.9)* are essential where revisions occur to ensure there is a method of keeping everyone informed; the latest design must be clearly marked.

The actual decision as to whether a construction becomes registered for ISO 9001 or 9002 will depend on how much design work is to be carried out and of what kind. The guiding principle is that 9002 is probably appropriate for an end product of standard design. But if the design work is for bespoke products and the clients' requirements are specific, then 9001 is likely to be more appropriate. Consequently, architects, consulting engineers, specialist subcontractors (where design is necessary) and design-and-build contractors will usually register on that basis.

Clause 4.5: document and data control

Consistency usually requires controlled documentation; this ensures four things:

1. Every person knows what is an agreed document, i.e. procedure, and when it should be used.
2. Every person is operating to uniform documents.

3. It is always obvious when a document has been modified or revised.
4. Obsolete documents are removed immediately.

Clause 4.5 helps to ensure that an organisation implementing a quality system has anticipated how the four objectives will be achieved. There will need to be specific procedures, and the quality manager will usually have overall responsibility for the task.

It goes beyond internal documentation, e.g. documents for the quality system, and covers any external documentation the organisation will use. An example would be building regulations, codes of practice, health and safety legislation, manufacturers' guidelines for using products, even the ISO standard itself.

The word *data* is deliberate; the quality system need not be in written form. Computers could easily replace a paper system.

Clause 4.6: purchasing

Purchasing is relevant to construction firms, particularly contractors who rely on subcontractors and suppliers. Essentially the organisation must evaluate all those suppliers of materials it uses to complete the product. The objective is to ensure that what they provide is capable of complying with the customer's requirements. It is worth noting that the standard uses the term *subcontractor* to those outside the organisation. It thus covers both suppliers and subcontractors as we normally understand them. The term *supplier*, when used in ISO 9000, means the organisation which supplies the customer, and is implementing the quality system.

There are three elements to this clause:

1. *Evaluation of subcontractors (4.6.2)* ultimately means using only those organisations demonstrably able to meet the requirements of their contract, hence to satisfy the final customer. A quality system should be able to assess subcontractors before their engagement and to monitor their progress throughout a job.
2. *Purchasing data (4.6.3)* means there should be a standardised way of placing orders for goods and services. The emphasis is on ensuring that what is required will be supplied. Thus, having considered how best to specify what an organisation requires, the method should incorporate all requisite information. The final choice of format is up to the organisation. For instance, telephone ordering is acceptable, providing a record is kept, perhaps using a standard form.

   Ultimately, it comes down to the confidence you have in the ability of others. In the current world of business, particularly in construction, the more written proof that is available, the less room there is for dispute.
3. *Verification of purchased product (4.6.4)* if agreed in the contract, allows the client to check products that will be incorporated into the finished product at the suppliers premises. In construction there may be many instances where the client would choose this option, particularly for items which will be highly visible in the building.

Clause 4.7: control of customer-supplied product

Clause 4.7 is highly relevant to construction. It covers the situation where a client wants certain materials delivered well in advance of being fitted (i.e. clause 16 of

JCT 80). It could be reasonably argued that the site or building to be worked on fall into this category. There are three implications:

1. The products are only used in the client's building.
2. They are well looked after in order to prevent deterioration and loss.
3. There is verification of their fitness for purpose on arrival. If not, they should be rejected and the client informed.

Clause 4.8: product identification and traceability

Clause 4.8 deals with component identification by marking or by accompanying documentation.

Identification, which will assist traceability, primarily ensures that components for particular locations actually end up there. It is also important to ensure that similar components such as bricks do not get mixed up between different parts of a job. Identification also enables the oldest stock to be used before more recent deliveries.

Traceability is the ability to know the history of a components. Should there be any problem, it is possible to trace who supplied an item and when it was fitted. This will make it possible to get the supplier or fitter to rectify or replace. Unique products are easy to trace, but bricks, concrete, etc., are more difficult, so the quality management system should include records of the batches that were used in certain parts of the building.

Clause 4.9: process control

According to Brown, process control 'is often seen as irrelevant by non-engineering companies' (1993:123). Satisfying the customer will consist of doing certain things. In order to achieve them, processes will be going on. A process is simply where inputs are transformed into outputs. Processes therefore go on in every organisation.

Figure 5.2 shows it is essential for an organisation to consider how it will control its processes. The aim is to ensure the end result is what the customer wants. Performance measures are necessary to indicate that customers are being satisfied. According to Jackson and Ashton (1995a:64), 'process control lies at the heart of the quality assurance concept.'

The use of procedures for effective process management is regarded as the best way to manage a quality system. According to Munro-Faure, Munro-Faure and Bones (1993:30), using the 'process model' has three advantages:

1. It ensures the QMS recognises the fundamental processes within the business rather than a series of tasks.
2. By controlling these processes, the QMS will enable you to meet customer requirements at minimum cost.
3. Documenting processes, rather than individual tasks, minimises the number of documents within the QMS. This helps to ensure the system is user-friendly and will be a real basis for managing the business.

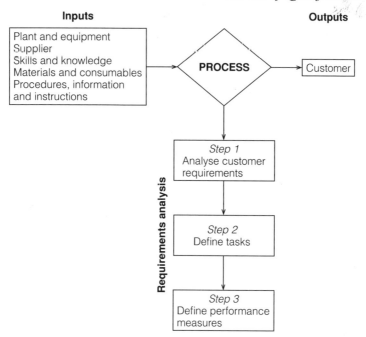

Fig. 5.2 Steps in process analysis (Adapted from Munro-Faure, Munro-Faure and Bones 1993:29)

The implication of process control in an organisation is that the appropriate resources and equipment are provided to ensure the processes are carried out consistently. This will include equipment, materials and people. All must be correctly maintained. In the case of people, this will require training (clause 4.18).

Clause 4.10: inspection and testing

Clause 4.10 considers three stages at which testing and inspection may be required:

1. Goods and materials upon receipt.
2. During the process of manufacture or construction.
3. On the final product before it passes to the customer.

Construction has a long tradition of carrying out tests. Concrete is an example of something that is tested on receipt (slump test). Subsequently it is tested for strength (cube tests in a laboratory). The Building Regulations specifically require tests for items like drainage. The contract usually specifies a defects liability period (DLP). It could be argued that QA should eliminate the need for this. Logically, having used ISO 9000, everything is right first time. However, tests and inspections remain an accepted part of the contract, and if something slips through, tests and inspection may pick it up.

The nature of the tests and inspections should be appropriate to the nature of the work. Some will be technical, i.e. strength characteristics. Others will be on appearance, a more subjective area. In construction, many items are covered over, so tests and inspections are a final opportunity to ensure any such work is correct and will perform as required. Problems which later manifest themselves will need much time and expense to uncover and rectify.

In practice there will usually be accepted methods and standard forms for tests and inspections. A quality management system will simply require them to be made into procedures which are followed consistently. It will be essential that all the forms are recorded and filed.

Clause 4.11: control of inspection, measuring and test equipment

This follows logically from clause 4.10. It requires records to be kept on all items of equipment that are routinely used to carry out inspections and tests. It is intended to ensure that any measurements will be accurate. Regular calibration checks should be documented and stored for future reference. Inspection equipment will vary from process to process and from industry to industry. A construction site will doubtless have theodolites, levels and tape measures. All of them should be accurate.

Clause 4.12: inspection and test status

It is sensible to have a method which clearly indicates the status of work, i.e. whether or not it has been checked. This allows identification of all work that has been passed, so it can move on to the next stage or be handed over to the customer. If work has not yet been inspected or tested, it should be impossible to use it by mistake or to pass it to the next stage.

The easiest way of doing this is to have markings, possibly coloured, on the relevant parts of the work. Where this is not possible, perhaps on finished brickwork, it will be best to have a marked-up drawing showing what has been checked, when and by whom. A matrix, together with colouring up of the completed sections, will achieve something which is clear and easy to interpret. This method was first witnessed on sites operated by a large contractor to provide records of foundations and drainage. It proved extremely successful, and adequately satisfied the requirements of ISO 9000.

Clause 4.13: control of non-conforming products

If, as a result of carrying out checks (tests and inspections), elements or products are found to be incorrect, two things need to be done:

1. Make it clear that this element or product is non-conforming.
2. Have procedures in order to decide what should be done, usually involving scrapping, repair or rework.

All instances will need to be recorded and documented.

Clause 4.14: corrective and preventative action

Clause 4.14 has two aims:

1. If a problem arises, it should be corrected at the earliest opportunity.
2. Problems should not be repeated; take steps to prevent reoccurrence.

Consequently, this section is dedicated to ensuring there are procedures to

- Identify problems
- Investigate their causes
- Provide solutions
- Prevent repetition

Clause 4.15: handling, storage, packaging, preservation and delivery

Clause 4.15 deals with procedures for goods which are produced but not sold immediately. Factory goods normally have to go through several stages before reaching the shop. It is essential to keep them in perfect condition, so as to avoid complaints about damage.

In the case of a building there is an onus on the contractor to avoid damage and deterioration of the components and finishes until they are no longer the contractor's responsibility. It is also prudent because any damage before handover will have to be rectified at the contractor's expense.

Contracts are usually very specific about how the contractor is to ensure the works are maintained (and insured) until they become the client's responsibility, i.e. until handover. It is therefore sensible to have procedures that deal with avoidance of accidental damage, theft and vandalism. For instance, if a part of the building is ready but not yet accepted by the client, there will need to be security, and possibly partial heating (to avoid frost in winter).

Clause 4.16: control of quality records

Clause 4.16 follows logically from the earlier clauses. If a quality management system is to be audited by a third party, records must have been compiled for comparison with the procedures. A quality record, according to Brown (1993:125), is something which demonstrates 'achievement of the required Quality and the effective operation of the Quality system.'

The quality records, like the quality system, should be designed to suit day-to-day needs. Procedures to control records should accommodate the feelings of those who compile them. It is both pointless and superfluous trying to make people implement an elaborate filing system for simple tasks; they will not see any point. As a result, the quality system will not work, and when the time comes for an audit, non-conformance will be more likely.

The other consideration is the time required to keep any records. Once again, this decision must be taken by the organisation that implements the quality system. A consideration may be the period to cover warranty, or perhaps the statute of limitations period. The use of microfiche and computer means that paper copies can be dispensed with after a set period, a situation which can incorporated into the system.

Clause 4.17: internal quality audits

The use of audits is perhaps where QA becomes controversial. Many regard auditing as a time to catch out the day-to-day users of the quality system. Auditing is thus perceived as a policing role, and is therefore resented. People normally do not like others checking on them. Unfortunately, auditing is the only way to discover how well the system is working.

The problem of auditing will be dealt with in the next chapter. However, the system must be credible. Initially there will be problems. Resentment and unwillingness to use procedures should be anticipated. Auditing can be made less confrontational by inviting users to write their own procedures. People can then see it is merely about confirming that normal practice actually occurs, practice now documented in a procedure. And whenever there is non-compliance, the procedures should be changed by the user, or the users should be challenged as to why they are not complying.

Clause 4.18: training

'The need for training to make the quality system work can be reiterated as "the quality of your output is only as good as the quality of your input". Input is not just materials, goods and documentation; it is also the skills, experience, knowledge, motivation and commitment of your personnel' (Wilton 1994:46).

Recognition that people are the most valuable asset in any organisation lies at the heart of many philosophies on quality management. People's ability to contribute is essential for smooth and efficient business operations. Without them, the potential to satisfy the customer will be seriously in doubt. An organisation which does not continually train its employees is at best foolish. It is not going to allow them to reach their full potential.

ISO 9000 requires three things:

1. Procedures must exist to identify training needs.
2. Records must be kept about the training of every employee.
3. Procedures must exist to satisfy the changing conditions within an efficient quality management system.

Training and education are at the heart of what many gurus expound; for instance, Deming makes it one of his fourteen points. This means they are an essential part of a quality management approach.

Clause 4.19: servicing

Clause 4.19 governs what happens after completion of the product or service. In the case of equipment, e.g. central heating, there will often be service periods incorporated into the contract. The main contractor will often be the link between the client and those carrying out the servicing. In other cases it may be rarer to offer servicing, although even professionals such as the quantity surveyor and designer may get queries after the job has finished.

The objective is for procedures to be written which cover how the organisation will organise, document and record servicing. The need to satisfy the client's requirements is, as always, paramount and it exists despite the end of the contractual period. As will be described later, devotion to customers is what differentiates the excellent organisations from those whose service is merely adequate.

Clause 4.20: statistical techniques

Clause 4.20 often causes concern. Construction tends to use statistics much more rarely than engineering. Dr Deming believed that improvement should be based on sound statistical data. His philosophy was based on the use of statistics to identify and solve problems. Without them, or indeed any other objective data, it will be difficult to know what needs to be improved.

Most organisations repeat a wide range of tasks over and over again. It makes sense to try improving these parts of the process to ensure the customer's needs are met. Consequently, if *simple* statistical methods are used, it should be possible to

1. Set targets.
2. Measure the achievement of progress against targets.
3. Identify how improvement can be made.

This falls into the territory of TQM, where statistical techniques are advocated as an essential improvement tool. ISO 9000 requires that, where statistics are used, there should be procedures to control the way in which they are applied and utilised. The objective is that people should use them because they assist in achieving their task. If, however, an organisation feels that the use of statistical techniques is wholly inappropriate, then it is better to state this in the quality manual than to force people to use a meaningless procedure.

SUMMARY

1. QA using ISO 9000 is a management system which seeks to achieve quality in the processes that an organisation uses to produce what its customers want.

2. There is no perfect way to implement QA. Any organisation should write a system which simultaneously suits its business objectives and complies with ISO 9000.

3. In addition to suiting an organisation's objectives, QA should not be 'forced' upon unwilling and uncommitted users. This will not be likely to produce benefit in the long term.

4. ISO 9000 contains a number of clauses to be followed when drafting an organisation's quality management system. Adherence to the quality management system, often dubbed the quality manual, is demonstrated using procedures that control tasks or operations which are routinely carried out.

5. The procedures which an organisation uses to control its quality management processes are verified against ISO 9000 by an external independent assessor (a third party). This is essential to maintain registration.

6. ISO 9000 has different versions to suit different operations. ISO 9001 and ISO 9002 are usually of greatest interest to the construction industry. ISO 9001 has twenty clauses, including a clause on design control; it is therefore relevant to design-only organisations and design-and-build organisations. ISO 9002 omits the clause on design control, so it is normally used by organisations that do no design work.

QUESTIONS

Discussion

1. How do you think organisations wishing to implement ISO 9000 should approach the task?

2. What reply would you give to those who say that QA is not relevant to what they do?

Individual

A small professional surveying practice has sought your guidance on how they might achieve ISO 9002. Write a report of up to eight pages summarising the main elements that would need to be considered, and flowchart a process for achieving a quality management system capable of accreditation.

FURTHER READING

Construction Industry Training Board (1990) *Guide to managing quality in construction*. CITB, Kings Lynn.

Jackson, P. and D. Ashton (1995) *Achieving BS EN ISO 9000*. Kogan Page, London.

McGoldrick, G. (1994) *The complete quality manual: a blueprint for producing your own quality system*. Pitman, London.

Munro-Faure, L., M. Munro-Faure and E. Bones (1993) *Achieving quality standards: A step-by-step guide to BS 5750/ISO 9000*. Pitman, London.

Waller, J., D. Allen and A. Burns (1993) *The quality management handbook: how to write and develop a successful manual for quality management systems*. Kogan Page, London.

Wilton, P. S. (1994) *The quality system development handbook, with ISO 9002*. Prentice Hall International, London.

CHAPTER 6

THE IMPORTANCE OF QUALITY MANAGERS

OBJECTIVES

- Explain the purpose behind QA (ISO 9000).
- Distinguish correct and incorrect implementation.
- Appreciate how QA managers encourage people's support.
- Anticipate resistance to the introduction of formal systems.

6.1   CONTEXT

Chapter 2 discussed the possible advantages and disadvantages of using a quality system such as ISO 9000. Any organisation wanting to introduce QA will want to ensure it achieves the advantages.

The most fundamental requirement for success is the right environment. Specifically, the system must be seen by users as being useful. If it is imposed, and consequently regarded as alien, and compliance is ensured by threats of punishment, then resistance and problems are likely.

This section deals with the way in which QA should be introduced so that potential problems are minimised. In so doing, it is necessary to draw attention to the importance of those who are charged with the job of achieving quality management. In the course of my own research, the task of introducing and managing QA normally falls to one person, the quality manager. This role is specifically required by ISO 9000 under section 4.1.2.3 (*management representative*).

Being a quality manager implies a role with great power. Indeed it can be. However, like any management role, there is a responsibility to use it carefully. Quality managers require support from those who use the quality system. Consequently, it is their efforts which will largely determine whether it is accepted or not. If the quality system is seen as adding value, then people will accept it. Alternatively, if the quality system is implemented without consultation and regarded as a bureaucratic imposition, then people will not accept it. The consequences of non-acceptance are that resentment builds up and the QA manager becomes an enforcer, insisting that procedures are obeyed.

6.2   STARTING ON THE RIGHT FOOT

The previous chapter contained my interpretation of ISO 9000; remember that others might have a different interpretation. The important thing about QA, using the quality standard ISO 9000, is that it is not meant to be a universal set of laws. It is not like the Building Regulations, for instance. It is to provide the major elements around which any organisation can design a system that suits its needs. As Brown argues (1993:13), 'There is a tendency to assume that a "system" must be something that is extremely elaborate and follows lots of clever rules and guidelines.'

The need to have procedures which cover all of the routine operations can seem daunting. Even though they do not have to be written, it is usually better to have them in that form. Nor indeed do they need to be lengthy. Brown continues: 'A system is whatever works for your firm [organisation] in order to process work smoothly and effectively from start to finish. The best system is one that has been proved to work. Most successful or well-established companies already have these in place in whole or in part, otherwise they would not be able to survive' (ibid.:14).

The objective is simply that existing 'good practice' is captured in the quality system. It is then important that the system is consistently used. However, for anyone coming to QA for the first time, it is sensible to ask the following pertinent questions:

- What needs to be done to achieve QA?
- What does this involve?
- Who writes the system?
- How can consistent use of the system be encouraged?
- Are there conditions which need to exist?
- What am I expected to do?

All are typical questions I encountered during the course of my research. Usually they were posed by people at the sharp end. QA for most firms was regarded as being essential in order to maintain registration. Senior management demanded that implementation occurred. Their answer to questions was typically of the sort: 'Our clients will not allow us to tender without it. We have to have it and therefore we *will* have it!'

Sometimes this meant that QA was implemented with little or no discussion. Any debate was viewed as resistance, and discouraged. The result was resentment. Senior management had failed to use the introduction of QA to produce something widely regarded as valuable.

In essence, the most important aspect of ISO 9000 is to have a workforce that feel the system is theirs. This is the task of whoever manages the system. But how will they manage it?

6.3   THE ROLE OF THE QUALITY MANAGER

Relatively little tends to be written about the work of a quality manager. This is perhaps surprising, given the importance of quality management. Jackson and

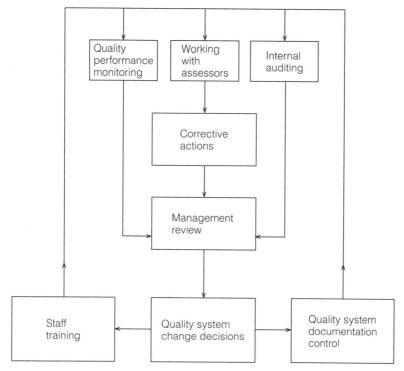

Fig. 6.1 The tasks of a quality manager (Reprinted with the permission of Kogan Page Ltd., from Jackson and Ashton 1995b:48)

Ashton (1995b) are exceptional in this regard. Figure 6.1 is their pictorial summary of a quality manager's tasks; they are described later in this chapter.

Contrary to popular misconception, the role of a quality manager is not exclusively technical and need not be executed in a highly mechanistic fashion. In practice the system must be used by people who work in the organisation, but as Jackson and Ashton admit, 'From the tasks indicated it may appear that the quality manager's role is relatively mechanistic, administration rather than management' (1995b:50). What they stress is that the role has two interrelated parts (ibid.:51):

- Administration (system skills)
- Management (people skills)

The administration skills consist of the following:

- Planning
- Record keeping
- Reporting
- Document production
- Liaison with assessors
- Understanding quality standards

These skills probably do require specialist technical know-how on what the organisation will need to do in order to achieve ISO 9000.

The management skills are much more about the need to manage human relations:

- Leadership
- Championing
- Facilitating
- Motivating
- Resourcing

It could be argued this is not a perfect list. For instance, it makes no explicit mention of communication. Never the less, it does contain the major elements of people management. The research I have carried out indicates that the success of a quality management system is demonstrated by people buying in. This success is therefore largely dependent upon the skill of the quality manager in convincing people of the need for the quality system. The quality manager must then be able to sympathise with any concerns, and support individuals' efforts.

Because quality managers require a variety of well-developed skills, they need to be carefully selected. Those who make the appointments should look beyond the short-term aim of getting the plaque on the wall. Although the immediate requirement is for someone to get the system up and running, the real objective is to find someone that will deliver long-term benefits. Such management skills are not exclusive to developing a quality management system. They apply to every level of management. However, they are the same characteristics that are particularly necessary when attempting other forms of quality management, such as TQM. Later chapters cover them in detail. Now it is time to look at some experiences of quality managers from construction companies. They more than amply demonstrate that, in conjunction with system skills, people skills are essential when implementing QA.

6.4   HOW NOT TO IMPLEMENT QA

This may seem like a negative way to describe the job of a quality manager. The research I have carried out indicates that a negative experience is not untypical of organisations who registered for BS 5750/ISO 9000 in the late 1980s and early 1990s.

The appointment of a quality manager, frequently from within, was quite often the first tangible step by an organisation towards introducing QA. The quality manager's task was usually put in very simple terms: they had merely to achieve registration. 'Read the standard and write something that complies,' that's what they were told.

It was not uncommon to be given permission to seek help from consultants. The government had financed the availability of grants to assist companies. In the rush to become registered, the need to have a system, any system, was regarded as the only important objective. The result was often too many procedures that were altogether too complex. The word *bureaucracy* is often associated with QA. This is not surprising. George, a QA manager, describes what happened:

We did what we were told by our consultant and if we weren't sure about something, you put it into the system. This meant that everything had a procedure. The trouble was the quality manual was like a telephone directory – and about as appealing. The amount of paper in circulation multiplied to the extent that people on site spent more time dealing with QA than their proper job.

This is something that Jackson and Ashton describe:

The charge of bureaucracy, is often very largely the consequence of a poorly designed system; it is needlessly bureaucratic because unnecessary form filling has been built into it or the particular form of record keeping is onerous because it is not well thought out. (1995a:42)

The reasons are often twofold:

- Insufficient user involvement from the outset
- Too much consultant involvement

This means the quality system is regarded as an additional burden on those who it should assist. The need to comply with lengthy, even irrelevant, procedures does not engender widespread support. Moreover, enforced compliance, in order to demonstrate the procedures work, only adds fuel to the fire. The quality manager will inevitably be caught between the senior managers, who stress that QA is essential, and those below who resent an imposed system. Keith, a QA manager, told me: 'You end up feeling that both sides are out to get you. It becomes a thankless task. I often say to myself there must be a better way.' There is.

6.5 WRITING PROCEDURES INTERACTIVELY

'A well designed system should run itself or, more accurately, be operated by staff without special management responsibilities for it' (Jackson and Ashton 1995b:45). The key phrase in this statement is 'well designed'. Once the system has been implemented, it must be accepted by the users. As a result, the QA manager's job should be easier. Rather than having to 'police' the system to ensure that procedures are being consistently followed, the quality manager can assume they are being adhered to and concentrate on facilitating improvement. This frequently enables the development of TQM.

So how can you aim to provide something that is well designed? For a start, avoid the approach of Section 6.4. Make sure the workforce owns the system. The procedures must be relevant to the tasks. This entails time and effort by the QA manager and by those who will actually use the procedures. Jackson and Ashton suggest 'it is essential to recognise that there is a distinction between responsibility for managing the system and responsibility for the system' (1995b:46).

Mike is a QA manager and he puts it like this:

'The system and procedures must reflect what really goes on. This is not as straightforward as it might appear. I've worked on site and know that there are some dodgy ways of doing things. Fair enough its supposed to be my job to eradicate them, but how? I can't do it by arm twisting. That would completely alienate them and confirm the suspicions of some that I've sold out when I took this job.

The trick is in getting people to accept that the procedures are best practice. As such they are not there to force people to do things in a way which is alien. No, on the contrary, it's a question of getting those who carry out tasks to ask themselves, 'Is the way it is currently done the most efficient?' In many cases, the current method is the best. But quite often you will hear the reply, 'That's the way it has always been done.' I then get everyone who is involved to think about alternative ways which might be better. It is often at this point that you hear things like, 'I've believed for a long time we could do it better, but no one asked me!'

This is usually the catalyst for others to join in. From this point it becomes a matter of getting agreement on one accepted way – not always easy. But the essential part is that the 'new' procedure has come from the users, not someone who is from outside. Therefore the rule for someone in this job is never to say 'Do it this way because I think it is better.' They have to make that decision themselves.

This was an approach which ensured that procedures represented a consensus view by those who would use them. Many stressed how this is a long and painful process. Change does not come easily to many. In construction there appears to be a belief that everyone does it their way. It takes a lot of courage to challenge the accepted ways of doing things. Consequently, it is important to choose someone who has the confidence to challenge accepted ideas.

6.6   MAINTAINING MOMENTUM

Getting the system up and running is the first hurdle for a quality manager. But what needs to be done in order to keep it going? The way the system is set up will govern the way it needs to be managed on a day-to-day basis. Jackson and Ashton believe the system should be operated with existing staff. However, they stress the need for someone to keep it maintained, otherwise 'entropy' will set in – the tendency for any system to become disorganised.

They advise that, if this situation occurs, there is a tendency for things to go wrong:

1. Defects in the system will not be rectified.
2. The aim of the system – to be consistent – will become discredited.
3. It will not be coordinated.
4. There will not be a central point for collecting data about how the system operates.
5. It is less likely the system will be internally audited.

Item 5 is usually the starting-point for external audits. If any of these items begin to occur, the whole effort to get QA will be regarded as a sham. It will then be extremely difficult to convince people that continued use of ISO 9000 is beneficial. Unfortunately, for many of the QA managers I interviewed, getting registration is regarded as the conclusion, not the beginning. Bill gave a typical description:

It is very easy to fall into the temptation of thinking that once you have the registration the job is done. The directors will come along and say well done. I've seen that some companies will invite the press and throw a party, like they've won an award.

And he stressed:

There's a danger in everyone thinking that the hard work is over, we can afford to relax. What I keep telling both those above and below, is that we must strive to use the system consistently, and in so doing get benefit. This means continually adjusting it so that we become better at our business.

I compare it to getting an MOT certificate on your car. All that this demonstrates is that it passed on the day. That your vehicle complies with the minimum standard. In between it can be a hazard, and you always run the risk of being caught by the police. The aim should be to have something, whether it be a car or ISO 9000, that is better than the minimum.

One of the biggest concerns of those I interviewed was the perception that QA was simply about doing enough to keep registration. Getting people to use the system consistently is not always easy. Don's experience as a QA manager taught him this:

You find that some people are not using parts of the system. So you ask them why. Sometimes it's forgetfulness. With others its laziness or couldn't be bothered. With both of these a stiff talking to will usually suffice. The really hard nut to crack is the one who says, 'Why should I have to comply? My work is fine – who's complained?' It is often the case that they had no part in drafting the procedure and have a gripe. My approach is then to get them to suggest how they believe it could be improved. You shouldn't be surprised if there is a torrent of abuse about QA reducing their expertise, and being a paper exercise.

Perhaps there is a temptation to let the resisters do their own thing. And perhaps it seems like an easy option. But Don does not believe it's an effective option:

You might as well give up. Everyone will see this as the green light to going back to their own individual systems. They will all give their reasons why they don't need to comply. It's the golden rule, no one is exempt.

6.7 OTHER DUTIES OF THE QUALITY MANAGER

Figure 6.1 indicates that a quality manager has eight areas of concern. Each area is an integral part of the total job of managing QA. Without covering all eight areas, it is unlikely the system will operate to its fullest potential.

6.7.1 Staff training

ISO 9000 contains a section dedicated to training (clause 4.18). This is to ensure that tasks are being carried out by competent employees who have received the appropriate training.

- **Introduction**
  - The organisation's unique quality system
  - ISO 9000
  - All staff are involved
- **Why a quality system?**
  - The benefits
  - Many other companies have ISO 9000
  - Dealing with negative preconceptions
- **The documented system**
  - The main elements:
    - Quality policy
    - Quality manual
    - Procedure manual
  - What is controlled documentation?
- **Procedures are mandatory**
  - Know and follow relevant procedures
  - But procedures can be changed
  - The progress of change is controlled
- **Change and quality improvement mechanisms**
  - Monitoring
  - Corrective actions
  - Management reviews
  - No disciplinary devices
- **Auditing**
  - Why are audits carried out?
  - What do auditors do?
  - Who belongs to the audit team?
- **Assessment**
  - Who are assessors?

Fig. 6.2   Motivational training agenda for quality management (Reprinted with the permission of Kogan Page Ltd., adapted from Jackson and Ashton 1995b:59)

However, before implementing the system, it will be desirable that every person appreciates what is involved and why. The main objective is consistency of operation. Even though QA requires nothing more than documenting what already happens, training will be beneficial. When the system starts to be used, people will begin to ask about their responsibilities:

- What procedures should they use?
- How should they be used?
- Which forms, checking and records are needed?

The subsequent seven areas will deal with the major concerns that are likely to arise. The most contentious is the need to carry out audits, either internal or external (see below). It will be advisable that as many people as possible are trained to carry out internal audits. This is sensible because it shares the quality manager's workload. It also means the quality manager does not have to take all the flak. Others more intimate with the operational side can explain this is a vital task.

A meeting with all employees, or perhaps sections in a large organisation, will give an opportunity to sell the system. Jackson and Ashton describe this as 'motivational training' (1995b:58). Figure 6.2 shows the sort of agenda they propose.

The guiding principle is *interaction* to facilitate discussion. This will aim to create a willingness to use the quality system, rather than telling people it must be used.

Even at this point there will probably be some who criticise the use of QA. It is vital to deal with their concerns. Any problem that arises will simply justify the views of detractors.

6.7.2 Quality system documentation control

Documentation control is the crux of QA. As Chapter 5 described, many sections of ISO 9000 are specifically dedicated to ensuring the quality system documents the following items:

- Tasks to achieve customer requirements
- Procedures for executing a series of tasks
- Compliance of procedures within a quality system

The QA manager will be accountable. Having overseen a system's introduction, they will take responsibility for all its aspects:

- Quality policy
- Manuals
- Procedures
- Records

The objectives remain consistency and uniformity, not always easy if each person is accustomed to doing their own thing. QA will require that, in future, people recognise how the system becomes the way to do everything.

A QA manager must ensure that all employees are complying as intended, usually by audit (see below). They must also keep the system updated and communicate any changes to those concerned. Everyone needs to be clear which is the controlled version of the quality system. The procedures are the most heavily used part of the system, so it is advisable that all users know which is the current version. A simple method to achieve this is by numbering or dating revisions to procedures.

6.7.3 Monitoring performance

The job of a QA manager demands continuous effort to keep the system running efficiently. The aim is to provide something that enables, rather than disables, people to carry out their tasks efficiently and with minimum interference.

Auditing will help to reveal problems with the system. It is then incumbent on the QA manager to understand the reasons for non–conformance. Where problems arise, the QA manager should go to the people concerned and attempt appropriate alterations to the system.

It is dangerous to wait for audits to uncover any problems. This is very much a reactive stance, and the consequences of failure may be costly, particularly if a client becomes affected. A better approach is for the QA manager to get those using the system to monitor its performance continuously. Users should be encouraged to improve their own procedures in a proactive way. A different style of management will be necessary to encourage a cultural change; problem solving needs to occur before the problems occur, not afterwards. This is a philosophy which may lead onto TQM (Chapter 7).

6.7.4   Audits

Internal quality audits are required under clause 4.17 of ISO 9000. They have several objectives:

1. To ensure the system is operating as designed.
2. To ensure QA is helping to meet client requirements.
3. To ensure the system's procedures are followed consistently.
4. To discover and explain any problems.
5. To make improvements to the system by combining items 1 to 4.

Figure 6.3 illustrates the audit process. Audits monitor the performance of a system, so it is very important to carry them out effectively. They must be seen as something

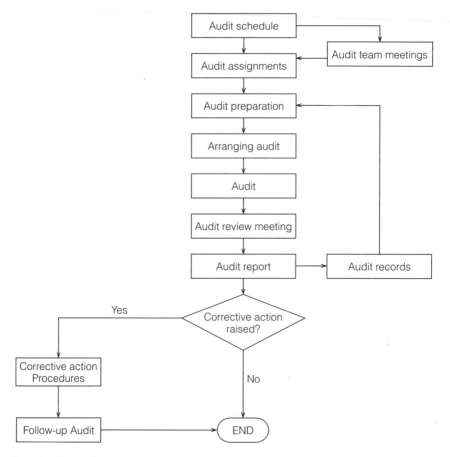

Fig. 6.3   The audit process (Reprinted with the permission of Kogan Page Ltd., from Jackson and Ashton 1995b:97)

that supports the efforts of those using the system. If, however, audits are regarded as policing – catching people out – it will hardly be surprising if they become resented. This will give internal auditors an unenviable task, to say the least. It is also likely to spell the end of constructive two-way communication, so it becomes extremely difficult to improve the system.

The QA manager will coordinate internal audits. Unless the organisation is very small, one person will not be able to do all the audits; it will be necessary to train some assistants. There is a strong argument for doing this: getting others involved will help them to realise what QA is about. It may also be that fresh eyes are able to see potential problems, hitherto unnoticed. Trained auditors should also provide a reserve during any period when the QA manager is away. At one of the companies I visited, the quality manager was taken ill suddenly. After many months he returned to find the system had largely been abandoned. As he explained: 'No one bothered to do any audits. This is something that I had always done. The fact that there were no checks meant everyone simply gave up doing QA. No one saw any incentive to continue.'

6.7.5 Working with the assessors

Many organisations find difficulty with the ISO 9000 requirement that their system should be assessed by outsiders, third-party accredited bodies. This assessment is crucial to obtaining registration. There will be no alternative, but careful management can alleviate potential problems:

1. It ensures that every person is aware of what the assessor is looking at – compliance with the system, not the actual quality of the work.
2. It ensures that visits by assessors are convenient.
3. It ensures the visits are seen as productive, not as a witch hunt. Non-compliances may be due to problems with the system, not incompetence or mistakes.
4. It ensures the assessor appreciates the process of construction.

The job of the QA manager will be to liaise with the selected assessor. An organisation is entirely free to choose who carries out the external audits.* In practice the person responsible for the quality system, the QA manager, will be most intimately involved, so the choice of assessor should be strongly influenced by them.

The official beginning of the assessor's task will start from the point when registration is sought. It can be terminated after registration, and another assessor appointed. It is probably desirable that the same assessor is used for a period long enough to bed the system in. Consequently, some unofficial enquiries are in order to establish the reputation of potential assessors. It will be useful to know how much assistance they are able to provide, particularly in the establishment of the system.

A good working relationship will go a long way to sorting out misunderstandings. The aim is to make their input something to be valued, not resented as outside

* Assessors usually belong to the National Accreditation Council for Certification Bodies (NACCB). A list of members can be obtained on request.

interference. As with any outsider, their ability to understand what goes on and why, will go a long way towards avoiding this perception.

6.7.6   Sorting out problems

It might be that an organisation's QA system was so well thought out, developed and implemented that everyone involved does exactly what is required and no problems occur. Although highly desirable, it is also unlikely. It should be no surprise that problems are likely to occur; indeed they should be anticipated.

Clause 4.14 of ISO 9000 requires that there are procedures to deal with problems immediately (section 4.1.4.2 on corrective action) and to stop reoccurrence in the future (section 4.1.4.3 on preventive action).

Problems may arise from many sources: from internal audits, from suppliers and subcontractors, from customers, as a result of external assessment. The aim is to achieve a system with these problem-solving attributes:

- Problem definition
- Cause investigation
- Solution formulation
- Solution optimisation
- System modification

Several solutions may be formulated; the best one is chosen at the next stage. The whole process will need to be documented in order that the problem can be closed down. It will allow subsequent traceability by anyone, even though they were not involved. This may be helpful in assisting the formulation of solutions to unrelated problems.

Jackson and Ashton provide a summary of the QA manager's role in taking corrective action (1995b:146):

- Problem selection and gatekeeping
- Ensuring problems are adequately defined
- Setting a timetable
- Appointing investigators
- Responding to the investigator's recommendations
- System change or implementing other decisions
- Auditing

The QA manager will have a strong influence on how problem solving is approached. The aim is to avoid problems. But when they do occur, the objective should be to minimise immediate effects in the short term, and prevent them reccurring in the long term. It is essential for the QA manager to ensure that problems do not become part of the 'blame' culture. To use current parlance, problems are 'opportunities for improvement'. If there is a willingness to stand back and think about the reasons, and to employ problem-solving techniques, particularly in a proactive may, the organisation will be moving towards the development of TQM.

6.7.7 Management review

Management review is specifically required when an organisation is operating
ISO 9000 (section 4.1.3). Even though the QA manager may be regarded as the key
person, a review must involve all those managers whose operations are covered by
QA. It should be regarded as a logical consequence of the tasks described above.
According to Jackson and Ashton (1995b:173–74), management review has the
following purposes:

1. Review the working of the quality system.
2. Consider problems identified through the quality system.
3. Agree and authorise changes in the system.
4. Agree and authorise other changes which may be appropriate.

The significance of a review will depend on the amount of importance it is given by
the managers involved. There are several reasons why all senior managers must attend:

1. To understand the aims of QA, overall and in individual departments.
2. To ensure the system is helping workers in their sections.
3. To become aware of any problems occurring in their departments, problems
   manifested by non-conformance.
4. To ensure any changes that are recommended, or being made, will be accepted
   and effectively implemented.

The QA manager is the person who will be in the best position to facilitate this
process. They will be able to report on the current state of the system; Jackson and
Ashton provide a checklist (ibid.:177):

- Results of operational non-conformity reviews
- Results of supplier reviews or monitoring
- Review of customer complaints received and results of customer satisfaction
  monitoring
- Audits carried out and their results
- Results of external assessments
- Corrective actions raised, completed or underway
- Corrective action recommendations authorised by the quality manager
- Changes made to the quality system since the last review meeting
- Monitoring or measurement of the quality system benefits

ISO 9000 requires meetings to be held at 'defined intervals', so the decision is left
to individual managements. Much will depend on how well the system operates
after the initial period of implementation, and this will be governed by the need
to have senior managers in attendance. It is not something that should be left until
there is some spare time; spare time will rarely occur by chance. In Ford, when the
board of directors meet, quality is the number one topic for discussion. Ford's
commitment to quality has demonstrated what this attitude can achieve.

The most essential element of this process is contained in the word *commitment*.
Easy to say, it is often much harder to demonstrate. Later chapters describe how the

commitment to quality, some would say passion, of senior management is the key to success. Leadership ability is certainly axiomatic to TQM, but is no less important to QA. Consequently, the management review should never be used as an opportunity for collecting reasons to go back and criticise a particular team or department. The system belongs to its users, and their ability to contribute should always remain the primary objective.

6.7.8   Changing the system

Clause 4.5 of ISO 9000 is concerned with ensuring that procedures exist to facilitate changes in the quality management system. There are many reasons why the system may need to be changed:

1. If problems are found when carrying out the activities described above, and if the system proves to be wanting, it is logical to make some changes.
2. An organisation may alter its methods, so previous procedures may no longer be representative.
3. The standard itself may change; ISO 9000 could be rewritten.
4. Users may feel the current version could be improved.

The danger inherent in using QA is that once it has been implemented, often with much effort, people may be loath to change it. The QA manager may find it hard enough to get everyone working consistently to the first version. It is worth remembering why QA will have been implemented in the first place: to ensure the organisation meets customer requirements. The aim, therefore, is to strive for continuous improvement of the system.

If everyone is satisfied the current version of the QA system is able to achieve its initial objectives, there will be no good reason for altering it. However, there is a world of difference between this, and finding that it is mainly okay. The QA manager will need skill to get everyone to continuously and critically examine the system. Engendering within users a desire for change is a real sign of success.

In terms of implementing changes, the QA manager is the person who will be responsible. They will need to ensure that changes are agreed and authorised, usually through the management review meeting. It is accepted that an urgent change might have to be made before this meeting. The QA manager will then need to do several things:

1. Check the draft change with those concerned.
2. Consider any implications for other parts of the system.
3. Make it obvious to everyone that a change has occurred (using a numbering and/or date convention).
4. Inform all users of the system about the change.
5. Ensure that every person incorporates the new version into their manual, usually by asking people to produce their controlled copy during an audit.
6. Keep a record of all changes.
7. Audit the actual use of changes to assess their appropriateness.

SUMMARY

This chapter has described how the quality system will be organised, and the role of the QA manager. The QA manager is important for several reasons:

1. Their efforts will largely determine the success of the system. This can be gauged by the willingness with which the system and its procedures are accepted by users.

2. They must ensure the system reflects what goes on. Only those who carry out operations and day-to-day tasks can advise on this.

3. They should have technical expertise in developing the system. They should also be able to give advice to employees on what administrative arrangements need to be completed to ensure the organisation's system is in accordance with ISO 9000.

4. They must have people management skills. This is crucial in enabling users to be committed to the system as a matter of routine. It will be tested by audits, perhaps internally by trained personnel, but especially by external third-party assessors.

5. They will be involved in many other duties that are critical to the health of the quality system:

   - Training all staff in the appropriate skills.

   - Ensuring the documentation associated with the system is accurate.

   - Monitoring the effectiveness of the system by audits.

   - Cooperating with external bodies such as the third-party assessor.

   - Updating the system when required.

   - Keeping senior management informed and committed to the use of QA as a means of producing organisational improvement.

   - Looking to the future to make changes which ensure that customer requirements, and their fulfilment are the organisation's main priorities.

QUESTIONS

Discussion

Quality managers can often find themselves as pig in the middle. What sort of person, and type of approach would be best suited to ensuring that QA is accepted as something which will add value? How do you think resistance to QA by users can be dealt with constructively?

Individual

Write an essay of about 2000 words covering these two items:

(a) Discuss the view that the quality manager is too often regarded as the only person who is directly concerned with quality matters.

(b) How should the role of quality manager be set up to ensure it is seen as coordinating effort, instead of writing the system for users.

FURTHER READING

Griffith, A. (1990) *Quality assurance in building*. Macmillan, London.

Hoyle, D. (1994) *BS 5750: quality systems handbook*. Butterworth-Heinemann, Oxford.

Hughes, T. and T. Williams (1991) *Quality assurance: a framework to build on*. BSP Professional Books, Oxford.

Jackson, P. and D. Ashton (1995) *Managing a quality system using BS EN ISO 9000 (formerly BS 5750)*. Kogan Page, London.

Thomas, B. (1995) *The human dimension of quality*. McGraw-Hill, Maidenhead.

Warwood, S. J. (1993) *Role of the modern quality manager*, Technical Communications, Letchworth.

CHAPTER 7

FROM QA TO TQM: ACHIEVING A CHANGE IN CULTURE

OBJECTIVES

- Understand that the results of QA will serve as a basis for moving to TQM.
- Explain that a vital aspect of TQM is an appropriate culture.
- Appreciate that culture may be nebulous, but it can still be described, e.g. by Handy's typology.
- Understand what sort of conditions are conducive to TQM.
- Describe how an organisation may need to change its culture to attempt to create the 'right' conditions.
- Explain the two paradigms of change management, planned and emergent.
- Appreciate that emergent change is the most sympathetic method to encourage TQM.

7.1 CONTEXT

The last chapter should have demonstrated that having QA is a dynamic process; change being intrinsic to it. To achieve this much is commendable, especially in an industry such as construction, where change is often resisted. For many construction organisations, ISO 9000 may be regarded as adequate for their immediate needs.

However, for some organisations, in having achieved a proactive system, certain things may have started to occur, particularly in people's attitudes towards quality. This is sometimes called a cultural change. The result of this may mean that QA is no longer regarded as being enough. In essence, this will be the start of the transition from QA to TQM. This chapter will consider the issues it raises.

7.2 BUILDING ON QA

Organisations that wish to implement TQM need not have implemented ISO 9000. However, in order to successfully develop TQM, it will be extremely beneficial to have been able to put in place a disciplined and systematic method of achieving best

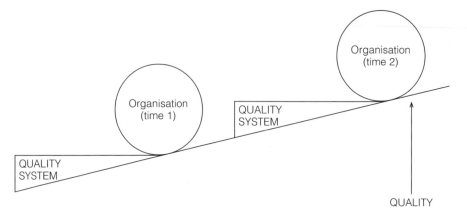

Fig. 7.1   The quality slope (Reprinted with the permission of Kogan Page Ltd., from Jackson and Ashton 1995b:205)

practice. Instead of being the way things should be done, the QA system will have become the way things are actually done. QA should be continuously monitored to ensure it represents the best way of achieving the desired results. It should therefore be used to assist in improvement. As Jackson and Ashton explain: 'Even if improvement is seen as the results of other sorts of activities, the quality system is still involved because it makes sure the rest of the processes continue to work effectively while change is considered and introduced in a particular area' (1995b:204).

Figure 7.1 shows how quality systems help to 'push' the organisation up the quality (improvement) slope. Indeed, as Jackson and Ashton advise, a QA manager who has achieved a successful quality management system can think about the next step: 'TQM is often thought of as a next step up from quality systems and a quality manager with an effective system in place can usefully think about some of the techniques involved' (ibid.:206). Others go further in warning of the danger in believing that implementation of ISO 9000 is the culmination of efforts towards quality. Brown is typical of those who hold this opinion: 'This is a short-sighted view which is likely to lead to disappointment and will not bring out the full benefits of a QMS' (1993:133). He firmly states: 'Instead ISO 9000 should be viewed merely as the first stage on the never ending road to Quality. The route to continuous improvement is the process commonly called TQM' (ibid.:133).

7.3   THE ROAD TO TQM

'Trying to explain TQM is difficult, as its methodology and end results are not as clear cut as ISO 9000, whereby a certificate awaits those who successfully stay the relatively short course' (Brown 1993:134). However, Dale, Boaden and Lascelles (1994:10) believe there are several key elements which characterise TQM. They advise that, if an organisation is attempting to move towards implementing TQM, it is necessary to address the organisational culture.

7.4 CHANGING THE CULTURE

Earlier in the book I described the philosophy of Japanese management. Their approach has been largely based on the statistical techniques which were developed by Americans such as Deming and Juran. The Japanese implemented them and demonstrated what could be achieved. Some believe that, in the rush to learn what the Japanese achieved, TQM has come to be seen as 'preeminently a Japanese innovation' (Burnes 1996:172).

It is believed that the success of Japanese industry is because of what Crainer calls their 'magpie like taking of the best of Western ideas and reinterpreting them to fit their own culture' (1996:140). The challenge that has been set for Western organisations is to 'relearn' the ideas that the Japanese have developed. As Pascale and Athos (1982) argued, the key difference that Japanese organisations tend to exhibit, is skill in the soft areas of management, i.e. people. This was essentially the main lesson that Peters and Waterman provided in their influential book *In search of excellence*, published in 1982.

Once people realised how important it was to address the needs of other people, culture became the crucial element of TQM and perhaps the biggest challenge to organisations wishing to improve. As Deming explained, appropriate use of the right hard systems such as SPC is all very well, but it's the people who really matter. Advocates of TQM believe it is crucial to address organisational culture.

What is actually meant by the term *organisational culture*? It will be useful to define it before considering the way in which it is supposed to be improved. But this is not easy. The reason, according to Brown (1995:5), is that no consensus exists about what organisational culture actually means. The passage people cite most often is rather long and maybe inaccessible:

> The culture of an organization refers to the unique configuration of norms, values, beliefs, ways of behaving and so on that characterise the manner in which groups and individuals combine to get things done. The distinctiveness of a particular organization is intimately bound up with its history and the character-building effects of past decisions and past leaders. It is manifested in the folkways, mores and the ideology to which members defer, as well as in the strategic choices made by the organization as a whole. (Eldridge and Crombie, 1974:89)

Perhaps this shorter definition is easier to digest:

> Culture is 'how things are done around here'. It is what is typical of the organization, the habits, the prevailing attitudes, the grown-up pattern of accepted and expected behaviour. (Drennan 1992:3)

Sadler makes the point that the culture of any organisation will probably have developed in an 'unplanned and unconscious way' (1995:65). As a result, he argues, it is often quite complex to trace the roots of the culture that exists. He provides a number of factors that may have influenced an organisation's culture (ibid.:65–66):

- Whereabouts the organisation began
- How the organisation developed
- Who are the senior managers
- What the organisation actually does
- How the organisation governs it members day-to-day

Perhaps you have noticed how certain people stick together. This is often because they share common views and interests. If there is enough interest, people frequently form a club or society to discuss things or pursue an activity. This situation is typical in any organisation, even in business. Kotter and Heskett (1992) studied corporate culture and its effect on business performance. According to their findings, there are several ways in which those who control an organisation ensure its culture is perpetuated:

1. Only choosing people who are known to be of a similar view to themselves – the people who fit.
2. Ostracising those who do not wish to conform.
3. The use of stories of past behaviour which have a powerful influence.
4. Continually drumming their values and beliefs into employees.
5. Rewarding those who demonstrate behaviour consistent with the accepted culture.

However, according to Peters (1992), Kanter, Stein and Jick (1992) and Handy (1994), what is really important in terms of the modern competitive world is to ensure that an organisation's culture matches the environment in which it operates. As Watson describes, 'What brings the activities of the organisational members to focus upon those purposes which lead to effective performance is the existence of a strong and clearly articulated culture' (1986:66). The challenge, therefore, is for those who wish to introduce a change in culture to facilitate the introduction of TQM, to ensure that certain things happen:

1. Everyone is encouraged to be concerned with their own work, and to dedicate themselves to continuous improvement.
2. The concept of the customer is engendered; that means internal and especially external customers.
3. Suppliers and subcontractors are treated as essential contributors to the process of improvement.
4. Honesty is encouraged in relationships and people naturally trust others to do a good job; this may eventually mean that QA procedures become superfluous.
5. Mistakes are accepted as part of the learning and improvement process; a blame culture ceases to exist.

7.5   CAN ORGANISATIONAL CULTURE BE DESCRIBED?

How is it possible to provide a starting-point for the change process, without at least being able to describe what presently exists, and, what is the most appropriate organisational culture to facilitate TQM? These questions are difficult to answer.

However, Handy, who believes there are four main types of organisational culture, provides a useful framework for considering change (1976:186–96). He describes these different cultures in terms of the organisational structures in which they typically exist:

1. *Power culture* locates control in one person or a small group. It is often associated with a web because, like a spider, the control is at the centre. According to Sadler, power culture can often be dynamic and it can also achieve success. He states that success will continue only 'as long as the source of power exercises sound judgment, makes the right decision, behaves with integrity and wins loyalty from others' (1995:71). A power culture is unlikely to encourage the sort of environment which allows TQM to develop. Unless of course the power holders are prepared to relinquish control.
2. *Person culture* consists of individuals who exercise professional independence; any structure will be pretty minimal. Person culture is often associated with clusters of stars, so it is questionable whether TQM is appropriate. If the individuals deal with a single customer or a few customers, there should be no real problems. But if the organisation wishes to become larger, or make its administration more consistent, then serious difficulties are likely.
3. *Role culture*, also called bureaucratic culture, places the onus on stability, rigidity and a mechanistic structure. There will usually be narrow job descriptions and a high attachment to procedures, rules, hierarchical position and authority. The focus of these organisations will be on predictability and security. The members of the organisation will follow orders; they will certainly be discouraged from acting on initiative. They will need to adhere to the organisational structure, and communication with other parts of the organisation will be exclusively through line managers. TQM simply would not work in this culture.
4. *Task culture* concentrates on getting the job done. Although there may be some administrative rules, they will be non-mandatory unless they help to complete the task. The structure is typically organic. The aim is to be flexible and to foster teamwork. Flexibility and teamwork lead to creativity and fast response to customer needs. Power is widely distributed among the members. This is the type of culture where TQM is most likely to be successful. Indeed, if the objective of TQM is to provide improvement in what the customers receive, it is a culture which must be encouraged.

7.6 HOW TO FACILITATE CULTURAL CHANGE

Culture is not something that comes in a box and can be plugged in like a computer system. It requires a number of things to be considered. Watson has a highly telling phrase, 'cultures are human-made' (1994:21). As he suggests, 'Cultures provide powerful guidelines for human action as well as resources to help us shape and justify actions which promote or defend our interests' (ibid.:21). The significance is that if the members of an organisation do not believe the view espoused by senior managers, they will provide their own. The key to success is for managers to aim for conditions which encourage all employees to share their view of the way that things need to be done.

Burnes (1996:116), in citing Dobson (1988), provides a four-step approach:

1. Consider ways to change the composition of the workforce so its members are more likely to share the views and 'values' which are desired.
2. Put into positions of influence those managers and workers who are committed to the shared views.
3. Make sure values and beliefs are adequately communicated. Burnes specifically refers to quality circles as a means to assist this process.
4. Change the formal systems of managing (particularly procedures) so they are sympathetic to the desired views and values.

Cummings and Huse (1989:428–30) provide what they see as five crucial steps for senior managers to facilitate cultural change:

1. *A clear strategic vision* provides everyone with an unambiguous understanding of the organisation's aims.
2. *Commitment* requires that the desired culture is something which those in positions of authority fully support. Lip-servers will impress no one, and worse, they will indicate there is division at the highest level. This is hardly likely to encourage others to follow.
3. *Symbolic leadership* is a manifestation of item 2. Actions speak louder than words, and this is undoubtedly true of cultural change.
4. *Support for organisational change* may be demonstrated by creating an appropriate organisational structure, with allied systems to enable the change to occur.
5. *Membership* essentially repeats Dobson's four-step approach. Cummings and Huse explicitly recognise that resistance is likely. They believe that resisters may well undermine any efforts to change the culture, so they should not be allowed to continue their employment, especially if they are senior managers.

7.7 PLANNED AND EMERGENT APPROACHES TO CHANGE

Cultural change may also bring problems (Salaman 1979; Schein, E. H. 1985). Some are concerned that, in creating the desired culture, a more resistant culture may also be implanted, perhaps making further change even harder. Besides, as Uttal (1983) suggests, the whole process is lengthy, taking anything up to fifteen years. Moreover, the belief that senior managers are able to influence cultural change is open to question. This is examined below. Two paradigms (models) exist: planned change and emergent change. Planned change recommends a top-down, rational approach; emergent change occurs from the bottom up.

7.7.1 Planned approaches to cultural change

Crucially, planned change views the process as linear. It involves moving from one step to another in a series of preplanned and predictable steps. Those in control, senior management, will make the decision to move towards a culture they believe is desirable. Consequently, they will initiate a programme of change management using a model to help them achieve their objective. One of the most significant assumptions of planned change is that 'an organization exists in different states at different times and that planned movement can occur from one state to another' (Cummings and Huse 1989:51). Three models are particularly associated with planned change.

Action research

According to French and Bell, this is 'research on action with the goal of making action more effective. . . . [It] refers to programs and interventions designed to solve a problem or improve a condition' (1984:98). This model is a rational and systematic approach where the initial emphasis is on considering the current situation. The group expected to produce the change will need to carry out analysis and develop a range of solutions. It is up to them, in conjunction with management, to implement the solution that appears most suitable. Its impact will also be evaluated so that learning can take place. The solution should prove to be more acceptable than the current situation because it has been arrived at collaboratively. Indeed, bringing people together will be useful in itself as it will help to establish shared values and beliefs. The desired culture comes as a by-product.

The three-step model

Lewin (1958) provided this model in order that the 'new level' (desired culture) is incorporated into the approach. Lewin believed this was necessary because, unless permanence is built in, those being changed will revert to past behaviour. In this approach the old way of doing things will have been discarded. The three steps are

- Unfreezing
- Moving
- Refreezing

This model is based on the need to convince people that their present way of working is incorrect and needs to be changed. Once this has been agreed, it is a matter of finding a new way to achieve the desired output. As part of this process, management need to set up systems and policies which support the effort.

Phases of planned change

This is a distillation of derivative models that stem from Lewin's original approach. Produced by Bullock and Batten (1985), it maps change in two dimensions:

- *Change phases* are the states through which an organisation moves.
- *Change processes* are the methods of change available to an organisation as it moves through its change phases.

Bullock and Batten's model identifies four change phases:

1. *Exploration.* A decision is made to consider the need for change and the desire to search for alternatives.
   Processes: be aware of this need; appoint change managers; plan the way in which the change will actually occur.
2. *Planning.* The actual change management is put in place.
   Processes: collect any information necessary to diagnose current problems; make decisions and implement them.
3. *Action.* Decisions are put into operation.
   Processes: make arrangements for the management of change to be achieved; allow evaluation and feedback.
4. *Integration.* This occurs once the change programme has been completed and implemented. The objective is to ensure that any changes to the organisation are accepted as 'the way things are done around here now.'
   Process: reinforce the required behaviour, values and beliefs so the new culture becomes the norm.

The planned models of change have been in existence for over fifty years, (Lewin first publicised his beliefs in 1946). They have thus provided the dominant theoretical method for managers wishing to implement change in organisations. However, there has been increasing criticism of these models for their prescriptiveness, as well as their reliance on rationality. There is also concern that planned change relies too heavily on the managers of organisations to produce the 'right solution'. The result is that, in an increasingly chaotic world, these models are regarded as inappropriate (Garvin 1993; Nonaka 1988).

7.7.2   The emergent approach to cultural change

Concerns over the planned approach have led to an alternative. The emergent approach is based on the belief that change is an open-ended and continuous process of adaptation. It primarily operates on the basis that cultural change is something that comes from an organisation's members. Consequently, it is best achieved by being bottom-up. As a result, emergent change is sympathetic with the central principle of TQM, continuous process improvement. Burnes puts it like this: 'The rationale for the emergent approach stems from the belief that change should not be and cannot be solidified, or seen as a series of linear events within a given period of time; instead it is viewed as a continuous process' (1996:187).

According to Dunphy and Stace, emergent change is necessary because of the need to have 'optimum fit' with the external environment (1993:905). Thus the key to success for an organisation is to ensure that its members are aware of changes in the environment. Benjamin and Mabey stress that 'while the primary stimulus for change remains those forces in the external environment, the primary motivator for how change is accomplished resides with the people within the organisation' (1993:181). The key aspect of emergent change is that senior managers are no longer the leaders of change. Instead their role becomes one of providing the members with the requisite skills and ability to recognise the changes in the environment.

Emergent change is a more recent approach than planned change, and it lacks a long history of models and techniques. Indeed, as Pettigrew and Whipp believe, there are no rules or models which provide the formula (1993:6). Rather, they suggest, the emphasis is on 'linking action by people at all levels of the business.'

In order to provide the cultural change where the efforts of those in the organisation are matched to the external conditions, three other features must be considered:

1. *Structure* will determine the way that people relate to one another. It should be equally sympathetic to formal and informal aspects. There is a strong link with 'culture-excellence', advocated by Tom Peters. The implication is that organisations are more customer-centred. Customers may be external or internal. Primarily, the objective of the organisation is to use emergent change to respond faster to shifts in the market. As Burnes suggests, the appropriate structure is a flatter structure with 'greater emphasis on effective horizontal processes' (1996:190).
2. *Facilitative management* means that managers seek to bring together different parts of the organisation to share problems and develop change initiatives. Managers must be prepared to be cope with uncertainty and take advice from those whom they have traditionally controlled. A key part of this new role is to communicate using both formal and informal channels. Managers should be involved in developing cooperation and consensus, but they needn't do this by leading a team.
3. *Organisational learning* assumes that all people within the organisation are able to contribute to making the organisation better able to respond to changing markets. However, in order to do this, they need to be able to experience and learn from the processes of change. As a consequence, non-managers should be encouraged to challenge established norms and practices. The result is that change becomes a bottom-up process, and managers must respond in a way which values these contributions. Pettigrew and Whipp believe that 'collective learning' is essential to create a sustainable change in culture (1993:18).

SUMMARY

This chapter has described what is required for an organisation to attempt the transition from QA to TQM. In order to achieve TQM, it is essential that certain aspects are considered:

1. The environment should encourage teamworking, communication and cooperation.
2. Creating these conditions, commonly called the organisational culture, will require analysis of the current state. This will then provide a starting-point for any change that is necessary.
3. If cultural change is required in an organisation, it needs to be facilitated by key personnel.
4. There are two ways of managing change; emergent (bottom-up) management is more likely to be successful than planned (top-down) management.

QUESTIONS

Discussion

Managing organisational culture is not something that can be achieved using the traditional technical approach. Discuss this statement in the context of what must be done to create an environment which facilitates TQM.

Individual

Think of an organisation you have dealt with recently, either as a customer or as an employee. Now write a report of up to eight pages that covers the following items:

(a) Describe its defining characteristics.

(b) What sort of cultural change do you think would improve its capability, and how might that change be achieved?

(c) If no changes are required (unlikely), what makes this organisation so good.

FURTHER READING

Anthony, P. (1994) *Managing culture*. Open University Press, Buckingham.

Atkinson, P. E. (1990) *Creating culture change: the key to successful Total Quality Management*, IFS Publications, Bedford.

Brown, A. (1995) *Organisational culture*. Pitman, London.

Burnes, B. (1996) *Managing change: a strategic approach to organisational dynamics*. Pitman, London.

Clarke, L. (1994) *The essence of change*. Prentice Hall, Hemel Hempstead.

Eccles, T. (1994) *Succeeding with change: implementing action-driven strategies*. McGraw-Hill, Maidenhead.

Fombrum, C. J. (1994) *Leading corporate change: how the world's foremost companies are launching revolutionary change*. McGraw-Hill, New York.

Mabey, C. and B. Mayon-White (eds) (1993) *Managing change*. 2nd edn, Open University/ Paul Chapman, London.

McCalman, J. and R. A. Paton (1992) *Change management: a guide to effective implementation*. Paul Chapman, London.

Quirke, B. (1995) *Communicating change*. McGraw-Hill, Maidenhead.

CHAPTER 8

THE IMPORTANCE OF PEOPLE

OBJECTIVES

- Appreciate that people are crucial to the success of quality initiatives which lead to organisational improvement.
- Explain the nature of motivation and how it may be used to encourage people during organisational change.
- Review the established theories of motivation and analyse their relevance to quality management.
- Describe Theory Z and other contemporary thinking on motivation.
- Explain the nature of empowerment and how it can be facilitated.

8.1 CONTEXT

Many quality improvements will have a technical aspect. There are systems of administration, means of communication and items of equipment. But all of them will be of little use unless they are made to operate with people. The Japanese developed *Kanban* to work alongside *Poka-yoke* and *Jidohka* (automation), but they have still to develop an entirely human-free method of production. For some this may be a Utopian vision, but for most it conjures up images from Orwell's *Nineteen eighty-four*.

Work for most of us is an economic necessity. But equally we are assumed to derive as much enjoyment as possible. Any quality effort should be based on the principle that if the people involved are convinced of the importance of improvement, they are more likely to be *motivated* to achieving it.

Motivation has attracted a great deal of interest in the last fifty years. There is debate as to how useful theories of motivation actually are. However, quality management initiatives require people to believe in the need for improvement. They must feel inclined to give their best. They must feel they are working for organisations which value their contribution. As a result, they will feel motivated.

8.2 A REVIEW OF MOTIVATION THEORY

Much has been written about motivation. As Chapter 3 described, motivation in the early factories was regarded as simply a matter of following the 'scientific laws' that Frederick Taylor laid down. According to Taylor, workers seek to maximise earnings; they do not care how they achieve it. Accordingly, managers 'must lay down in detail what each worker should do, step by step; [and] ensure through close supervision that the instructions are adhered to; and to give positive motivation, link pay to performance' (Burnes 1996:28).

The birth of a more compassionate view of workers' motivation came from the Hawthorne studies. There appears to be a clear connection between the finding of the Hawthorne experiments and the work of Shewhart, Deming and Juran. Deming and Juran placed great emphasis on workers being given the chance to work to the best of their ability. Even more significant, managers cannot delegate their responsibilities. Managers have a duty to provide a system of operations that will allow workers to achieve their best (remember Deming believed that managers are responsible for over 90% of the problems).

Despite the work at Hawthorne being carried out in the 1920s, it was not until the 1950s that more investigation was carried out into motivation of workers. Three individuals in particular are particularly associated with the development of what has become known as the human relations approach: Douglas McGregor, Abraham Maslow and Frederick Herzberg.

8.2.1 McGregor's Theory X and Theory Y

Theory X is a simple assumption that workers are lazy, and require to be treated in a Tayloristic way. Theory Y is a more considerate view that workers are capable, and given the right encouragement, they will develop their own initiative and become self-motivated. McGregor (1960) suggested that X-type organisations would contain people who were unable to innovate in response to changing circumstances. Consequently, he advocated that organisations should move towards a more consensual approach which encouraged the ingenuity, imagination and creativity of workers; they should become Y-type.

McGregor has much in common with the advocates of TQM and William Ouchi provides an obvious connection. Before his death in 1964, McGregor was working on a new theory that brought together the needs of the corporation and the individual. He called it Theory Z. Almost two decades later, William Ouchi chose this as the title of his book. Ouchi used Theory Z as a way to compare differences between the way that Japanese and American organisations motivate their workers. Theory Z is described in more detail later in this chapter.

8.2.2 Maslow's hierarchy of needs

Maslow (1943) developed a hierarchical model (usually represented by a pyramid) of what he believed are people's needs (Figure 8.1). The principle is that any human

Fig. 8.1 Maslow's hierarchy of needs (based on Maslow 1954)

will start by satisfying basic needs. Once they have met the basics, they will try to satisfy other needs, gradually moving up the pyramid.

Physiological needs – food water, shelter – are the lowest level. Once they are satisfied, the individual will move onto the next level of the hierarchy, safety and security. This level provides protection from both physical and emotional harm. Once they are satisfied, the individual will move onto the next level of the hierarchy, love and belongingness, and onto self-esteem. Esteem needs have two dimensions, internal which is self-respect and personal achievement, and external, which is perceived status and recognition. The highest level is self-actualization. This predicts that every person will try to work to the limits of their ability. Unless they do, they will feel they are underachieving.

Maslow argued that managers should identify their workers on the hierarchy. Once they have done this, they should help (motivate) them to move to the next level. In so doing, workers will contribute to both the organisational, and personal wellbeing. The importance of the hierarchy is that 'it explained why in some situations Tayloristic incentives were effective, while in other situations, such as the Hawthorne Experiments, other factors proved more important' (Burnes 1996:51).

As Oakland points out (1993:323), it will be hard for people to feel motivated in a situation where they feel threatened, perhaps when redundancies are likely. 'Driving out fear' was one of Deming's fourteen points. The Japanese have, until recently, made lifetime employment one of the principles upon which organisations operate.

Despite Maslow's theory having the virtue of being simple, neither Maslow nor other researchers have been able to provide empirical evidence to validate it.

8.2.3 Herzberg's two-factor theory

Herzberg (1968) identified two types of influences that lead either to satisfaction or dissatisfaction (Figure 8.2). He called the satisfaction influences motivators and the dissatisfaction influences hygiene factors. According to this theory, management should ensure the hygiene factors are adequate to be comparable with those of competitors. Thus, if your competitors are producing high quality output and

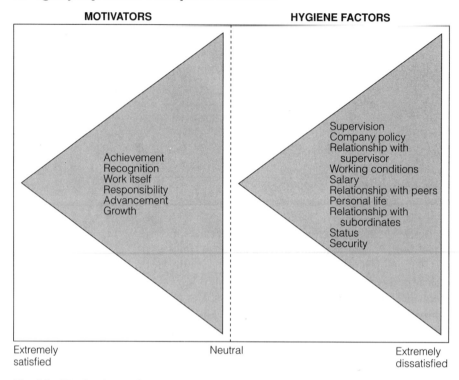

Fig. 8.2   Herzberg's two-factor theory (Robbins 1994:468)

provide better facilities, then that is the first thing to address. What Herzberg particularly stressed was that if you want to get your workers to produce more, or better, then the motivating factors need to be considered. Motivating factors are associated with the job and must allow the worker to exercise more input or control over the task.

The two factors are essential in any effort to improve quality, and motivators are very much in line with thinking on quality management. In particular, Herzberg advocated what he called job enrichment, whereby workers are given increasing control over the work they do. Decisions are taken by individuals or semi-autonomous teams; previously they would have been taken by managers. The Japanese have used this concept to good effect.

8.2.4   Content theories

Maslow's theory is a content theory; it tries to predict the needs that people have. It provides the direction and conditions which assist workers to fulfil their needs, and it has implications for organisations implementing quality management initiatives for improvement. Other content theories are Alderfer's ERG theory and McClelland's socially acquired needs theory.

Alderfer's ERG theory

This theory was propounded as a refinement of Maslow's hierarchy of needs. The letters of the acronym come from three needs which Clayton P. Alderfer (1972) identified as important:

- Existence
- Relatedness
- Growth

Alderfer differed from Maslow in that he believed an individual may be influenced by all of these needs simultaneously. Also, an organisation should be aware of its workers' needs.

What everyone must work to achieve is customer satisfaction. Failure to address its customers' needs could cause an organisation to go out of business, threatening the security of its members. If all the members, managers and workers, share the same objectives, and strive towards improvement, there will be relatedness. Finally, growth will result from people being satisfied they have provided output which is their best. The improvement should lead to the goods or services being more able to compete. As a result, and in line with Deming's belief, more business is created – the organisation will grow.

McClelland's socially acquired needs theory

Unlike Maslow and Alderfer, David McClelland (1961) did not believe that needs were inherent in people. He suggested that needs can be acquired in different social situations, and he proposed a list of three:

- Achievement
- Affiliation
- Power

The importance of this theory is that people are given tasks which match their developed needs. In an organisation attempting to introduce improvement, different people will have different needs. In addition, the needs of each person may change throughout the quality management initiative, as it tries to develop greater input and awareness.

8.2.5 Process theories

Content theories consider internal forces (needs) that are believed to motivate; process theories try to predict how people choose behaviours to attain these needs.

Reinforcement theory

Reinforcement theory predicts that people determine their behaviour in accordance with the perceived consequences. It therefore suggests that people ask, What's in it for me? when they consider their input and behaviour.

Organisations should therefore aim to provide conditions which attempt to encourage (reinforce) positive effort by everyone. The most widely recognised proponent of reinforcement was Skinner (1953). He suggested that managers should use what he called operant conditioning. This assumes that people learn to regulate their behaviour as a result of experience. In particular, they are believed to avoid consequences which are likely to be 'painful'.

Managers should therefore use reinforcement by arranging selective rewards and punishments. Rachlin (1970) believes there are four methods by which managers may attempt to modify behaviour:

1. *Positive reinforcement* means that rewards are available to those who perform well.
2. *Avoidance learning* means that something perceived as unpleasant is removed following desirable behaviour.
3. *Punishment* is a 'painful' or unpleasant consequence available to be used when behaviour is undesirable.
4. *Extinction* is the withdrawal of all rewards in the expectation that behaviour will improve; after sufficient improvement, rewards are reinstated.

Reinforcement is an area where there may be a strong temptation to use rewards and punishments to get people to comply with expected standards of behaviour in order to develop a TQ 'culture'. However, Deming would take a dim view of this method. In particular, he would criticise the use of fear and rewards, punishment and performance pay.

Vroom's expectancy theory

This is somewhat similar to reinforcement in that it tries to explain and predict how people rationalise their behaviour in order to derive certain outcomes. People are believed to pursue certain patterns in order to produce an expected outcome. Victor Vroom (1964) postulated that expectancy could be summarised in the following formula:

Motivational force = valence × instrumentality × expectancy

Where the three components on the right-hand side are defined as follows:

- *Valence* is the satisfaction an individual attaches to an outcome.
- *Instrumentality* is the calculated link made between performance and certain behaviours or inputs.
- *Expectancy* is the probability that certain behaviours will produce successful performance.

Bounds, Dobbins, and Fowler (1995:442–43) suggest that expectancy theory has certain implications for managers implementing quality management:

1. Managers should develop shared understanding with employees of the values of certain behaviours. In particular, it is important that employees see the importance of their efforts in giving customers high quality. This is more likely to result in job security and higher wages.

2. The shared sense of importance in achieving the organisational goal of giving customers high quality should be underpinned by the use of open and honest communication and cultural change.
3. Feedback processes should be created so that everyone can perceive a clear connection between what they are doing and the accomplishment of organisational goals.
4. Training and education programmes should be set up to give employees the ability and confidence to make the judgments associated with expectancy.

They make the very valid point that if people do not understand, appreciate or have the knowledge to believe in something, they will be 'very likely to avoid it.'

Vroom's theory has attracted criticism because of its apparent assumption that everyone acts in such a calculating way. Some (de Charms 1968; Deci 1975, 1976) have stressed how the use of extrinsic motivators, i.e. those designed to please others, may adversely effect a person's own intrinsic motivation. However, organisations which have achieved recognition as being 'high quality producers' (Xerox and Motorola) use extrinsic rewards.

Goal-setting theory

This is the belief that having goals will help people to focus their energy and provide a shared commitment to achieving them (Locke 1968). Four factors increase their likelihood of attainment:

1. They should be ambitious but reasonably achievable.
2. They should be set by those who will work towards them (at all levels).
3. They should be specific and measurable.
4. They should represent long-term improvement, not just short-term gain.

There is concern that goal setting, although very useful for monitoring progress, can become the main obsession of managers. Deming warned against the use of numerical quotas. In particular, he condemned the use of target setting by managers. He thought that targets were dangerous because they could be misused, ultimately doing more harm than good. Managers should concentrate instead on improving the overall system, leaving the targets to others.

Social comparison and equity theory

Adams (1963) put forward the theory of equity. It is based on the way in which people compare their behaviour and performance relative to others. Adams believed that people look for fairness and consistency in treatment. If they believe this has occurred, they are likely to be satisfied. They will perceive the existence of equity.

But if people perceive inequity – perhaps they see unfair or inconsistent treatment – they are likely to feel unhappy. The result is demotivation. Furthermore, they are likely to take corrective actions:

1. Change the level of their input, usually a reduction.
2. Alter the outcome by asking for more money, or in extreme circumstances by stealing to 'restore' equity.
3. Practise self-deception by convincing themselves the inequity is not really important.
4. Change the reference source by comparing their situation with something that superficially diminishes the inequity.
5. Leave and go to another organisation.

Bounds, Dobbins and Fowler (1995:450) believe the principles of TQM are consistent with equity theory. The importance of everyone acting as a team is essential to quality management initiatives. Inequity will undermine any improvement effort.

8.3 MORE CONTEMPORARY THEORIES OF MOTIVATION

Not surprisingly, given the recent efforts to use quality management as a means to create improvement, established theories of motivation are being re-evaluated. Existing theories may still be considered by those wishing to implement quality management. However, more contemporary and directly related theories have been proposed.

8.3.1 Theory Z

One way to ensure that people are motivated during efforts to improve quality was provided by William Ouchi in his book, *Theory Z* (1981). Theory Z was a model which Ouchi proposed to American firms in order to emulate the quality standards achieved by Japanese producers.

Ouchi (1981), who had been influenced by McGregor's work on Theory X and Y (Section 8.2.1), was impressed by Chris Argyris' book, *Integrating the individual and the organization* (1964). In particular, he concurred with Argyris' view that the challenge to managers was to 'integrate individuals into organizations, not to create alienating, hostile and impersonally bureaucratic places' (Ouchi 1981:83).

Ouchi's Theory Z stressed that workers are more likely to be motivated when the following characteristics exist:

- Hierarchy that does not seek to constrain
- Lack of bureaucracy
- Consistency in shared objectives

As Ouchi stressed, Z-type organisation are 'most aptly described as clans in that they are intimate associations of people engaged in economic activity but tied together through a variety of bonds' (ibid.:83). Ouchi believes that workers need to be involved in making decisions. This, he stresses, requires openness and cooperation.

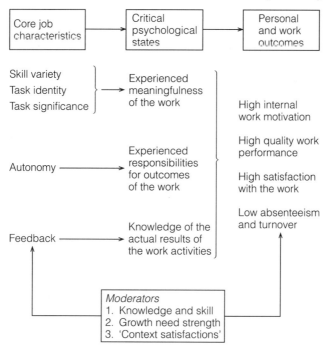

Fig. 8.3 A job characteristics model of work motivation (Adapted from Hackman, J. and G. Oldham (1980) *Work Redesign*, p. 90, figure 4.6. © 1980 by Addison-Wesley Publishing Co., Inc. Reprinted by permission of Addison Wesley Longman Inc.)

8.3.2 The job characteristics model

The job characteristics model (Figure 8.3) was proposed by Richard Hackman and Greg Oldham in 1976. Their model was in fact an elaboration on what Hackman had produced in 1971 with Ed Lawler. Based on research into the psychological effects of work, five factors are associated with motivation in a job:

1. *Skills variety* is the number of activities that a job requires.
2. *Task identity* is the extent to which a person is involved in the total process.
3. *Task significance* is the importance the individual attaches to their job in the context of the total process.
4. *Autonomy* is the amount of discretion allowed to the worker.
5. *Feedback* describes the extent to which the individual learns about the outcomes of their performance.

As Dale and Cooper suggest, this model stresses the need for managers to communicate to workers the importance of their efforts in achieving quality output (1992:84). They also believe that the model emphasises the need to give the worker responsibility and 'ownership' over their own task. In particular, they believe it provides justification for

- Learning and practising problem-solving skills used in quality improvement
- Use of teamwork
- Applying quality management tools and techniques
- Control of the whole process
- The need to formally audit by using ISO 9000

The importance of this model is that the worker is no longer seen as a simple device, to which managers apply motivational 'tricks' in order to get the required behaviour and performance. Workers at every level, especially the operational level, are given much greater autonomy and involvement. This is entirely consistent with the principles of TQM and a new word has been coined to describe it – empowerment.

8.4   WHAT IS EMPOWERMENT?

Robbins argues that in today's organisations many jobs are so complex that 'workers are more knowledgeable than their mangers' (1994:53). Thus methods associated with Taylorism, where managers tell people what to do and treat workers as 'unthinking machines' are no longer appropriate. On the contrary, Robbins believes that the contemporary manager needs to recognise that 'they can often improve quality, productivity and employee commitment by redesigning jobs and letting workers make job-related decisions' (ibid.:53). He calls this process empowering, and describes it as a more beneficial way to treat workers.

   Clutterbuck cites Richard Carver of the Coverdale Organisation to define empowerment as the method by which employees are encouraged to 'take personal responsibility for improving the way they do their jobs and contribute to the organisations goals' (1994:12). McBride and Clarke suggest there has been a 'decline in faith in traditional autocratic management' (1996:117). This, they explain, requires all employees to be empowered, especially those closely in contact with customers. And the result of empowerment 'expands the skills and tasks that make up a job, and gives team members and front-line workers greater control over decision making, problem solving, action planning, and goal setting, thereby liberating creativity and innovation' (ibid.:117–18).

   Crucially, they stress that, in order for an organisation to compete in terms of quality, it requires 'an empowered workforce [which] is an essential element of any Total Quality Management programme' (ibid.:117–18). Bounds, Dobbins and Fowler concur with this belief when they admit that, even though empowerment is a buzzword, in TQM it is necessary because 'it ensures that employees know their roles in implementing the organization's mission [towards improvement]' (1995:97).

8.5   WHAT DO GURUS SAY ON EMPOWERMENT?

Most writers on the subject of TQM tend to offer broadly similar suggestions for how managers can empower their workers. Empowerment may sound like a modern word, but it is not such a recent concept. Peter Drucker, regarded as the guru's guru (Micklethwait and Wooldridge, 1996:71), can claim to have identified empowerment in his book, *The concept of the corporation*, published in 1946. His description of 'creating the self-governing plant community' and identification of 'knowledge workers' can be seen as an early justification for what is now known as empowerment. His thinking laid down the foundations for many subsequent advocates of the concept, in particular Tom Peters, Charles Handy, Rosabeth Moss Kanter and Ricardo Semler.

Tom Peters' message since *In search of excellence* is for the need to focus on people and less on the structure of organisations. His belief is that the successful organisation tends to be unrestricted by formal structure and hierarchy. Instead, organisations should respond to rapidly changing markets by being dynamic. This, Peters advises, requires expertise, curiosity and innovation by 'people close to the action'. His central message, although written in increasingly esoteric language, can be found in books such as *A passion for excellence* (1985), jointly authored with Nancy Austin, *Thriving on chaos* (1989), *Liberation management* (1992), *The Tom Peters seminar: crazy times call for crazy organizations* (1994) and *The pursuit of wow! Every person's guide to topsy-turvy times* (1994).

Handy is a British-born writer who, whilst being as prolific as Peters, is not as much of an evangelist or self-publicist. He advocates that organisations need to change the way they operate. Handy believes that, in the future, organisations will be much smaller and consist of configurations he describes as shamrock, federal and triple I. Handy strongly recommends that pursuit of profit is not everything. He stresses how the skills of people must be incorporated so they have the ability to respond to rapidly changing markets.

His views are not confined to management. Many of his books deal with economics, politics, sociology and religion: *Understanding organizations* (1976), *Gods of management* (1978), *The future of work* (1984), *The making of managers* (1988), with J. Constable, *The Age of Unreason* (1989), *Inside organisations: 21 ideas for managers* (1990), *Waiting for the mountain to move and other reflections on life* (1991), *The empty raincoat* (1994) and *Beyond certainty: the changing world of organizations* (1995).

Kanter is known for two influential books: *The Change Masters* (1984) and *When giants learn to dance: mastering the challenges of strategy, management and careers in the 1990s* (1989). As Crainer points out, Kanter is largely responsible for the 'rise in interest – if not the practice – of empowerment' (1997:136). Burnes explains that Kanter thinks the future organisation will be one where employees are 'given greater freedom to innovate and experiment' (1996:87). Kanter particularly stresses that, in the future, organisations will operate in a knowledge-based society. These organisations will have to share power with employees at all levels.

Ricardo Semler is somewhat different form the previous three writers in that his beliefs are not theoretical. His views, contained in a book called *Maverick* (1993), are based upon experience of empowerment at his own organisation, Semco, which manufactures pumps and cooling units in Sao Paulo, Brazil. On his first day at Semco after taking over from his father, Semler dispensed with 60% of the top management. Since then, despite operating in a country known for economic turbulence and hyperinflation, Semco's profits have gone up fivefold, and productivity sevenfold.

Semco is based upon workplace democracy. Every employee is in effect a manager with access to all company information. Employees set their own wages and targets, and decide how profits should be reinvested or distributed. As part of the radical transformation he initiated, Semler's own job has even disappeared. The task of chief executive is a rotating one. Semler does not attend Semco on a daily basis, and when he does, his advice is not mandatory.

8.6   HOW CAN EMPLOYEES BE EMPOWERED?

Sirkin (1993:58) provides a list of five essential points to achieve empowered employees:

1. Senior managers should be 'put in the shoes of those to be empowered.' This, Sirkin believes, will give them 'empathy' and understanding of what people at the operational end 'need to do their jobs.'
2. Employees should be given authority. Without the ability to make decisions, and without support for their decisions, be they right or wrong, employees cannot fully contribute. As Sirkin advises, mistakes can be valuable in that they provide an 'opportunity to learn and improve.'
3. As part of giving authority to make decisions, employees need to feel comfortable. They will need training and education to develop the skills and confidence to exercise authority, authority which comes with empowerment.
4. Employees should be given all the necessary resources to solve problems. This will include the information that is germane to the task. It also includes the provision of time to consider problems. When the solution is presented, the physical resources must follow, e.g. plant, labour and finance. Senior managers cannot then override a decision from empowered employees because they don't like it, or believe a cheaper option is required.
5. Finally, Sirkin recognises that empowerment of employees will normally be a decision taken at senior management level. As he points out, this does not suit everyone, particularly middle managers whose authority is being relinquished. Middle managers 'must see that their jobs now involve managing employees' skill level, not their decisions.' As a result, middle managers are now more akin to being coaches, rather than a 'traditional controller of labour' that is responsible for everything.

McBride and Clark (1996:120) have listed the same points as Sirkin along with some extra ones:

- Employees need to be trusted; they also need to feel that managers have faith in them.
- A culture of interdependence should be created.

All of these points are well summarised by Clutterbuck (1994) in Table 8.1. Clutterbuck also provides a useful summary of the techniques for empowerment (Table 8.2).

Table 8.1 How to move from disempowerment to empowerment

Disempowering	Empowering
Fear	Challenge and adventure
Learning is a chore	Learning is an adventure
Dependence	Mutual independence
People take little initiative	People solve their own problems
	People suggest solutions to other people's problems
	People have the skills to work without supervision
Scant training and development	Continuous development
Avoiding change	Continuous change welcomed
Feedback is seen as criticism	Feedback seen as essential
Past experience has no relevance	Pause, reflect and learn
Training and development is the responsibility of personnel	Training and development is everyone's responsibility
Lack of vision	Strong, focused and shared vision
Problem avoiding	Problem solving
Closed communications	Open communications
	■ sharing of information
	■ sharing of ideas
	■ sharing of skills
Distrust and suspicion	Trust

Source: Reprinted with the permission of Kogan Page Ltd., from Clutterbuck (1994:52–53).

Table 8.2 Techniques for empowerment

	Manual workers	Clerical workers	Professional workers
Scope of knowledge-base	Job rotation	Job rotation	Job enlargement
	Group technology and self-governing teams	Self-governing teams	Narrowing focus to allow people to become experts
		Job enlargement	
Scope of discretion over ■ when tasks are done ■ how tasks are done ■ design of new tasks ■ interpretation of policy	Self-governing teams	Self-governing teams	Self-governing teams
	Problem-solving skills	Problem-solving skills	Problem-solving skills
	Delegation	Delegation	Delegation
Involvement in policy making	Referenda		
	Cascade briefings (with an emphasis on upward communications)		
	Participation (from consultative committees through employees on the board)		
Opportunity to change systems that affect people in other functions/departments	Suggestion schemes		
	Cross-functional quality improvement teams		

Source: Reprinted with the permission of Kogan Page Ltd., from Clutterbuck (1994:44).

8.7 ORGANISATIONAL CHANGES TO FACILITATE EMPOWERMENT

The main implication of empowerment is that an organisation becomes less hierarchical. It becomes flatter in order to facilitate communication and openness. Advocates of empowerment point to common criticisms of traditional hierarchical structures:

- Remoteness of senior managers
- The amount of bureaucracy
- Slow decision making and long response time
- Lack of cooperation

Lack of cooperation may occur between departments on the same level or on different levels of the hierarchy.

Clutterbuck (1994:60–62) provides five types of 'radical' structure which may assist in facilitating empowerment:

1. *Bull's-eye.* Customers are in the middle and the organisation is built around their needs.
2. *Amoeba.* The emphasis is on continual change in shape and division into small units as customers alter requirements or the amount of business expands.
3. *Star.* A few employees work very closely with customers. It is very much like project management teams who become almost part of the client's organisation.
4. *Boundaryless.* An extension of the star structure to make it 'so responsive that barriers between customer and supplier become so vague as to be meaningless.'
5. *Chemical soup.* Another manifestation of project management where teams come together for a specific project, 'then dissolve into new combinations as needed.' This is something that construction has been doing for a long time.

SUMMARY

People who work in organisations can be encouraged to contribute towards improvement using quality management. There are five important points:

1. Getting employees to actively support and be committed to the quality initiative is essential.

2. Theories of how workers should be treated have developed considerably since the advent of the factory system.

3. Many theories exist to explain motivation, but some of them have little relevance to quality initiatives such as TQM.

4. Contemporary theories such as empowerment, which explicitly encourages workers to take more responsibility and control over their work, are highly conducive to positive organisational change.

5. If empowerment is to be facilitated, the workers involved must be trained and educated in order that they can actively contribute.

QUESTIONS

Discussion

Consider the following statement:

> The construction industry is regarded as an industry where the treatment of workers leaves a lot to be desired.

(a) How true is it?

(b) What could the industry do in order to improve its capabilities using more positive treatment of workers?

Individual

Consider the trades that are normally used on a construction site. Write about 2000 words to analyse how the workers who carry out these trades should be used in order to achieve the desired results, i.e. completion on time, within budget and to the required standard. In particular, are bonus schemes still relevant today?

FURTHER READING

Bowman, C. and M. G. Jarrett (1996) *Management in practice*. 3rd edn, Butterworth-Heinemann, Oxford.

Clutterbuck, D. (1994) *The power of empowerment: release the hidden talents of your employees.* Kogan Page, London.

Dale, B. and C. Cooper (1992) *Total quality and human resources*. Blackwell, Oxford.

Hannagan, T. (1995) *Management: concepts and practices*. Pitman, London.

Langford, D., M. R. Hancock, R. Fellows and A. W. Gale (1995) *Human resources management in construction*. Longman, Harlow.

Lavender, S. (1996) *Management for the construction industry*. Addison Wesley Longman, Harlow.

McBride, J. and N. Clark (1996) *20 steps to better management*. BBC Books, London.

Oates, D. (1993) *Leadership: the art of delegation*. Century, London.

Tsoukas, H. (1994) *New thinking in organisational behaviour*. Butterworth-Heinemann, Oxford.

CHAPTER 9

LEADERSHIP FOR TOTAL QUALITY MANAGEMENT

OBJECTIVES

- Analyse the role of management in quality management initiatives.
- Understand what is commonly known as leadership.
- Examine the established theories of leadership and their relevance to quality management.
- Explain how ideas of leadership have come to embrace more radical approaches.
- Analyse the link between appropriate leadership and success.

9.1 CONTEXT

Chapter 6 described the tasks that a quality manager is required to perform when implementing ISO 9000. Quality managers do not appoint themselves. They are given the responsibility and authority to carry out their job by senior management. ISO 9000 is normally required as a minimum to tender for work being let by clients such as government agencies, and other public sector bodies. Thus, for many organisations, becoming registered to QA is a vital strategic decision.

Chapter 7 explained why there are good reasons to consider ISO 9000 as the starting point of the transition to TQM. Having chosen to go for TQM, two things are essential. The first and most important is the need for the senior managers to support the changes being undertaken and to provide the right sort of leadership.

The second is that the person who has day-to-day responsibility for facilitating the required cultural change needs to possess the skill and confidence necessary to achieve the task. This person, frequently the quality manager who has been responsible for the implementation of ISO 9000, will also have to provide the right sort of leadership.

9.2 LEADERSHIP AND TQM

According to the advocates of TQM, leadership is axiomatic. It is regarded as one of the most essential elements. Stahl is typical of those who promote TQM when he states:

Leadership is especially important in Total Quality Management because TQ involves dramatic change to a new and improved way of doing business and managing operations. It takes influential leaders to cause followers to change. (1995:302)

Others reinforce this belief. Bell, McBride and Wilson explain that:

Recent evidence indicates that organisations succeed or fail in their desire for positive management of quality in direct proportion to the amount of visible commitment from senior management level. (1994:189)

Dale and Cooper in no uncertain terms believe that:

Without the total commitment of the CEO [chief executive officer] and his or her immediate executives and other senior managers [presumably the quality manager in particular], nothing much will happen and anything that does will not be permanent. They must take personal charge and exercise forceful leadership. (1992:20)

But what is meant by the term *leadership*, and how can it be implemented to ensure a TQ initiative will be successful in any organisation?

9.3 THE MEANING OF LEADERSHIP

Attempting to define the term *leadership* is not easy. Indeed, Charles Handy (1976:93) has been moved to state that 'the search for the definitive solution to the leadership problem has proved to be another endless quest for the Holy Grail in organization theory.' Thomas (1993:109) goes even further, 'Although [leadership] is a subject that has attracted a great deal of attention and . . . has spawned a voluminous literature, it remains one of the most confused areas in the whole field of management.'

This does not bode well for understanding what leadership is, and how it can be applied to TQM. However, despite the problems, it is worthwhile to at least consider the available theories and what they might offer.

9.3.1 Qualities or traits

This is the belief that leaders are born with certain characteristics. It is these characteristics which give them the skills to be effective leaders. People with such traits are able to influence others to act in accordance with their wishes. Six key traits have been identified by Kirkpatrick and Locke (1991:48–60):

1. *Drive* makes a person ambitious and willing to seek new challenges.
2. *Motivation* makes a person want to lead others.
3. *Honesty* means a person is regarded as trustworthy and having integrity.
4. *Self-confidence* is expressed by a person in all that they do.

5. *Intelligence* allows a person to deal with varied problems and to absorb large
   amounts of information.
6. *Knowledge* allows a person to understand the organisation, its people, its market
   and how to ensure its output is appropriate for changing expectations.

In organisational terms, if there is the desire to implement TQM, the best people to
lead the initiative are those with similar traits. However, this is not as straightforward
as it might seem. Jennings (1961:2), points out that 'fifty years of study have failed
to produce one personality trait or set of qualities that can be used to discriminate
between leaders and non-leaders.'

9.3.2   Functional or group

This approach focuses on what the leader does (the functions) and how they
influence followers (the group). Advocates of this approach, such as Kotter (1990),
believe that once these functions have been identified, others can be trained to use
them.

Krech, Crutchfield and Ballachey (1962) identified fourteen functions:

- Executive (oversees all activities of groups within the organisation)
- Planner (plans how objectives are achieved in both the short term and the long
  term)
- Policy maker
- Expert (advises the group)
- External group representative
- Controller of internal relations
- Purveyor of rewards and punishments
- Arbitrator and mediator (controls conflict and disagreement)
- Exemplar (provides a role model for others to follow)
- Symbol of group (around which others gain focus and identity)
- Substitute for individual responsibility (insecure members of the group can rely
  on the leader)
- Ideologist (provides beliefs and sets standards)
- Father figure
- Scapegoat (anger can be directed at the leader in a crisis or during failure)

This long list illustrates the complexity of the different functions of leadership.
How they can be learned is obviously not going to be easy. But they do at least
emphasise the importance of the group. Indeed John Adair has developed the idea
of 'action-centred' leadership to assist in identifying and meeting group needs.
There are three elements to this approach:

- *Task functions* achieve the common task for the group.
- *Team functions* ensure the team is built, maintained and able to meet its
  objectives.
- *Individual functions* are needed to give those in the group confidence and
  recognition for their efforts.

Adair (1983) stresses that, in order to meet these three needs, the leader must be aware that not all will be equal, and their relative priorities may alter. The leader will need to understand this process and skilfully judge what is most important.

Adair's emphasis on the importance of teams is something the advocates of TQM strongly support. Chapter 10 examines the use of teams in TQM. Other contributors to leadership place great reliance on the use of teams.

9.3.3 Behavioural theories

Behavioural theories are concerned with the belief that what is important about a leader is their behaviour. Thus the aim is to discover those aspects of behaviour that make people good leaders. Once these behavioural characteristics have been established, it is assumed that anyone can be trained in their use.

Ohio state studies

This was a research project which involved the interviewing of over a thousand leaders and their subordinates (Fleishman 1974:1–6). Two behavioural characteristics proved to be important:

- *Consideration*: the amount of trust, respect and empathy which exists between the leader and their followers.
- *Initiating structure*: the way in which the leader structures (defines and organises) the task for their followers.

These two characteristics were used to form dimensions of a matrix against which a leader could score low or high. Thus a leader could be low/low, low/high, high/high, or high/low on consideration/structure. Early research indicated that the high/high leader was most likely to achieve best follower performance. However, it has been criticised as being too superficial to be of any practical use.

University of Michigan studies

Carried out at the same time as the Ohio studies, their objectives were similar. The findings (Kahn and Katz 1960) also indicated there were two dimensions, but this time they were 'employee-centred' and 'production-centred'. Kahn and Katz believe that, in order to achieve higher support from followers, a leader should be employee-centred, something with which advocates of TQM would concur. However, this study suffers from the same criticism as Ohio, superficiality.

The leadership grid

This was a development of the concern for people/concern for production scales, by having nine points of measurement on each. Originally proposed by Blake and Mouton (1985), it was further developed by Blake and McCanse (1991). A 9 × 9 grid provides 81 different combinations of leadership behaviour, although it is normal to consider only 5 (Figure 9.1). Not surprisingly it is the 9,9 combination

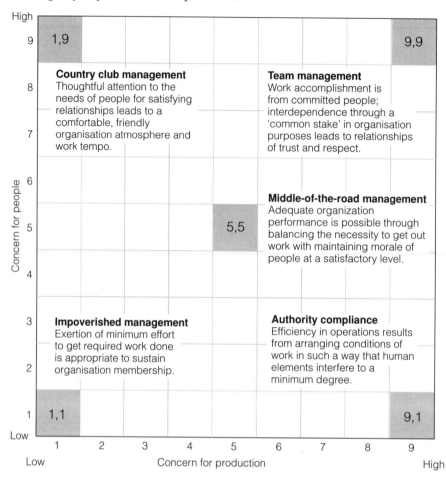

Fig. 9.1   The leadership grid (From *Leadership Dilemmas – Grid Solutions*, p. 29 by Robert R. Blake and Anne Adams McCanse. Copyright 1991 © by Robert R. Blake and the Estate of Jane S. Mouton. Used with permission. All rights reserved.)

(team) that is advocated as the state to which any leader should aspire. As Bounds, Dobbins and Fowler suggest, this is consistent with TQM because 'managers who practice team-orientated behaviors are those most likely to lead their employees in the harmonious and successful achievement of organizational goals' (1995:401).

However, they warn it is not easy to achieve. They suggest that the manager must 'integrate the inputs' of the team so the output represents a collective effort and uses each individual's talents. They suggest four points of action that a team leader should consider:

1. Get the members to input into goal setting when planning.
2. Know the strengths and weaknesses of each member.
3. Use the strengths of individuals to achieve outstanding performance by the group.
4. Continually train people to develop their abilities to contribute in the future.

Styles of leadership

This theory is based on the belief that the processes of interpersonal behaviour adopted by a leader will have a direct effect on the performance of followers. Mullins (1993:242) identifies three styles:

1. *Authoritarian/autocratic*, where the leader possesses all the power to make decisions, set policy, determine procedures and control rewards and punishments.
2. *Democratic*, where the group have equal power with the leader to make all the necessary decisions about how to achieve objectives.
3. *Genuine laissez-faire*, where the leader adopts a hands–off approach and exists merely to offer advice and guidance when requested.

The Continuum of leadership

The most widely recognised model to deal with leadership styles was provided by Tannenbaum and Schmidt (1973). Figure 9.2 shows how the leadership styles occupy a continuum. The two leadership extremes, boss-centred and subordinate-centred, are placed at opposite ends of the continuum. By moving from the left (boss-centred) towards the right (subordinate-centred), the leader will be able to choose from many different styles, but four main candidates emerge:

1. *Tells*, where the leader identifies what is to be done, chooses the way to do it and imposes their will.
2. *Sells*, where the leader will make the decision as to what to do, and how to do it, but will attempt to elicit support by persuading the group to agree with their intention.
3. *Consults*, where the leader identifies what is to be done, but will listen to advice from the group on how to achieve the objective. The leader will still maintain the right to choose the course of action.

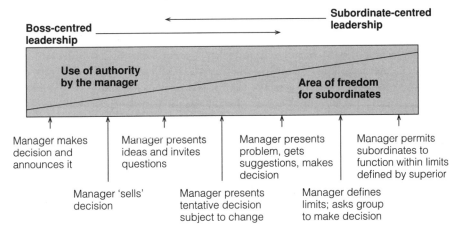

Fig. 9.2   The leadership continuum (Reprinted by permission of *Harvard Business Review*. An Exhibit from 'How to choose a leadership pattern' by Robert Tannenbaum and Warren H. Schmidt, *Harvard Business Review*, May–June 1973. Copyright © 1973 by the President and Fellows of Harvard College; all rights reserved.)

4. *Joins*, in which the leader as the manager is able to provide the benefit of their expertise, but is an equal member and is bound by what the group decides.

As Tannenbaum and Schmidt suggest, three forces will influence which type of style is most appropriate:

1. Forces in the leader, such as their confidence in themselves and the group members.
2. Forces in the individual group members, such as their ability and skills in terms of the complexity of problems, and their interest in developing personal solutions.
3. Forces in the situation, which means recognising the importance of the group, and the likely impact of the group's decision in terms of the organisation's ability to compete.

Situational forces are particularly important when the leader of a group wishes to ensure their contribution will assist the efforts of TQM. If every group in an organisation can individually improve, the collective improvements of the groups will improve the organisation as a whole.

9.3.4　Contingency theories

A manager who wants to adopt the most effective style of leadership will need to be aware of certain changes that may occur, particularly under external forces. Tannenbaum and Schmidt (1973) believe the most powerful and variable force to which managers must respond, comes from the external environment. This will greatly affect the situation to be dealt with by any manager or group.

It was on this assumption that a number of theories were developed. These theories are called contingency theories because their proponents believe that a leader's behaviour is highly dependent (contingent) on the situation in which they find themselves.

Fiedler's contingency model

The first comprehensive model of contingency was developed by Fred Fiedler (1967). Fiedler researched the proposition that the best leaders are those who have high concern for both production and people. His findings led him to believe that a manager cannot achieve high concern for both at the same time. Instead, he proposed that personality, in conjunction with experience and knowledge, has a strong bearing on how a person will behave, hence on how they will lead.

Fiedler's theory has two behavioural dimensions, task and relationship orientation. Leaders who are task orientated will be primarily interested in getting the job done, even at the expense of relationships with followers. A leader who is relationship orientated will also want to get the job done, but will ensure that good relationships are maintained with their followers.

According to Fielder's predictions, the success of a group is likely to be highest when there is a match between the leader's behavioural orientation and a combination of three situational factors:

1. Leader–member relationships.
2. Task structure, how clearly the task is defined.
3. Position power of the leader, how much formal authority is held by the leader.

Fiedler predicted that a favourable situation is one where the following features exist:

1. The leader is accepted and trusted (high leader–member relationships).
2. The tasks to be carried out by the group are clearly defined (high task structure).
3. The manager has high authority in their position (high position power).

An unfavourable situation is the converse of all these.

Fiedler carried out over 1200 studies to verify his theory. He used what is known as the least preferred scale (LPC); this is a questionnaire which purports to measure the rating that a manager or leader gives about the person with whom they could work least well. A high score from the LPC indicates relationship orientation. A low score indicates task orientation. By analysing the questionnaires, Fiedler was able to propose eight combinations of group–task situations which have a bearing on the appropriate leadership style (Figure 9.3).

Fiedler predicts that very favourable situations or very unfavourable situations (I, II, III, VII and VIII) require a task-orientated leader (low LPC score) with a directive and controlling style. But moderately favourable situations, where the variables are mixed (IV, V and VI), will be better suited to a relationship-orientated leader (high LPC score) with a more participative style.

This model has been criticised (Filley, House and Kerr 1976) on the basis of uncertainty in interpretation and the aspects of leadership behaviour it represents. And Rice (1978) believes that the scores given by respondents are not stable. However, the model does stress how different managers will be better suited to particular situations. In organisations attempting to adopt TQM, the emphasis is usually on teamwork. As a result, good leader–member relationships would be appropriate, so a manager will need to be relationship orientated.

There may still be situations which call for completion of the task as the priority. Then it may seem logical to use a task-orientated leader. However, in a TQM organisation where the culture is one of shared problem solving, a task-orientated leader will probably find it more beneficial to use tools and techniques to identify and resolve any impediments to completing the task.

Hersey–Blanchard situational theory

This is a model of situational leadership proposed by Paul Hersey and Kenneth Blanchard (1974). Essentially it asserts that a manager who wants to lead others must change their style in accordance with the needs of followers, and the tasks they are required to achieve. No matter how effective they think they are, any manager will be judged successful, or otherwise, on the basis of what their followers do.

Hersey and Blanchard use the term *maturity* to describe the willingness and ability of people to complete a task by directing themselves. There are two components:

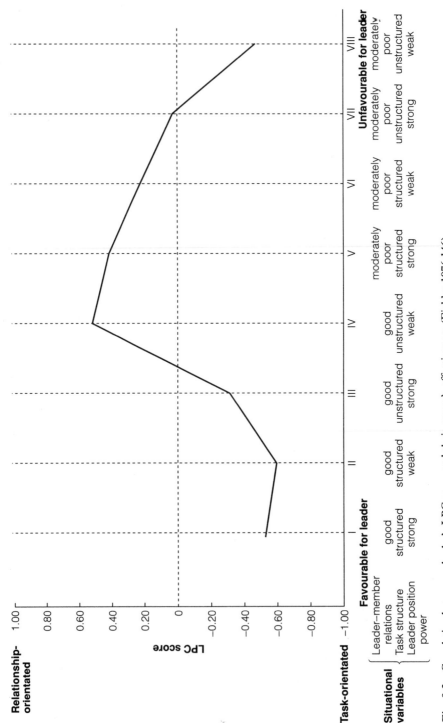

Fig. 9.3 Correlation between a leader's LPC score and their group's effectiveness (Fielder 1976:146)

- *Job maturity* is the knowledge, ability and expertise that someone uses to carry out a task without needing to be directed.
- *Psychological maturity* indicates the willingness or internal (intrinsic) motivation that a person possesses.

In the model that Hersey and Blanchard provide, there are four stages of maturity:

- *M1*: followers are unable and unwilling to take responsibility. They are neither committed nor motivated.
- *M2*: followers are unable to do the job but are willing to learn how. They are motivated but need to be trained.
- *M3*: followers have the appropriate skills and knowledge to do the job, but they are unwilling to work or they lack sufficient motivation.
- *M4*: followers are able and willing to do the job.

Hersey and Blanchard use the same dimensions as Fiedler, task and relationship behaviours. The crucial difference is that they classify them as high or low. These results are then combined with four specific leadership styles:

1. *Telling* (high task, low relationship) the emphasis is on guidance by the leader; most appropriate to M1.
2. *Selling* (high task, high relationship) the leader is both directive and supportive; most appropriate to M2.
3. *Participating* (low task, high relationship) the leader will engage in discussion with followers and share in decision making; the leader is a facilitator; most appropriate to M3.
4. *Delegating* (low task, low relationship) the leader needs to provide very little guidance or support; most appropriate to M4.

Hersey and Blanchard's model is shown in Figure 9.4. It seems to indicate that the more training, education and experience a worker receives, the more mature they become. This will mean they require less leadership, and ultimately no leadership. They will have become empowered. This change is desirable in a TQ organisation, where the emphasis is on encouraging workers to become more involved.

However, research carried out into this model has not been very encouraging. Vecchio (1987) found that evidence only partially supported it. Blank, Weitzel and Green (1990) found no evidence to support it at all. Nicholls (1985) felt that Hersey and Blanchard violated three principles: consistency, continuity and conformity. He provided a revised model to deal with these violations (Figure 9.5):

1. A group which is at development stage needs a leader who acts as a parent. They must encourage followers to increase their ability and willingness. As these qualities develop, the leader moves from telling, through consulting to developing.
2. The role of developer aims to encourage participation by increasing delegation.
3. If a follower's willingness develops more quickly than their ability, the leader should act as a coach.
4. But if a follower's ability develops in advance of their willingness, the leader should act as a driver, to ensure that performance is maximised.

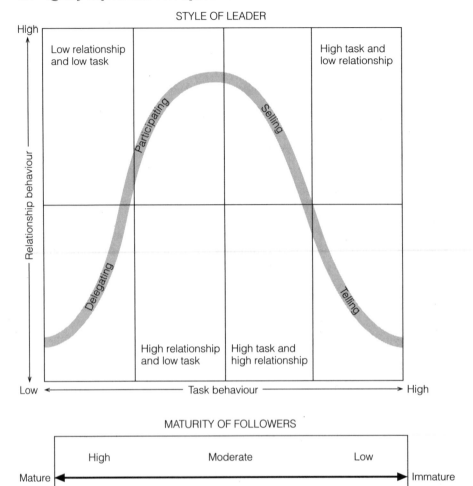

Fig. 9.4 The situational leadership model (Hersey and Blanchard 1982:152)

Path–goal theory

This is a theory developed by Robert House (1971); it proposes that a leader should assist their followers to attain their goals by providing the 'right' direction (path). The leader should also encourage them to have goals that are compatible with other group members, and the organisation in general. It indicates that workers have preconceived ideas about what levels of performance will lead to satisfying their goals. As such, it is similar to what is called the expectancy theory of motivation.

Path–goal theory suggests that the way in which the leader is accepted by followers will depend upon the leader's ability to assist them in satisfying their

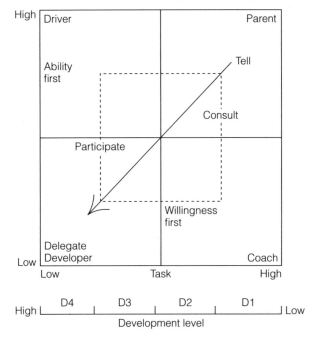

Fig. 9.5 The 'new' situational model (Reprinted with the permission of MCB University Press, from Nicholls 1985:60)

goals. House identified four types of leadership behaviour to help followers in satisfying their goals:

1. *Directive* leaders are specific about what is to be done and how.
2. *Supportive* leaders will be willing to give advice and guidance to followers in order to achieve goals; they are also concerned about followers' needs and welfare.
3. *Participative* leaders consult with followers before they decide on the goals.
4. *Achievement-orientated* leaders will set challenging goals for the group, and will expect it to perform at the highest level.

House believes these different behaviours can be employed by the same person at different times to suit varying situations. There are two determinants (situational factors) to the leader behaviour:

■ Followers' characteristics
■ Nature of the task

The implications of this approach are that managers who want to lead must get to know their followers very well. The leader must also be willing to be flexible. Both attributes indicate the need for someone to lead who is able to empathise with others. There is an assumption on building relationships and teams.

Leader–participation model

Devised by Victor Vroom and Phillip Yetton (1973), this model links leadership behaviour with decision making. It identifies three aspects:

1. Any decision should be related to organisational performance.
2. Any decision should be acceptable to the group that will implement it.
3. There should be enough time to make the decision properly.

Vroom and Yetton suggest there are five leadership behaviours which may be used to suit certain situations:

- *Autocratic (AI)* the leader solves the problem and decides how to implement the solution.
- *Autocratic (AII)* the leader obtains information from the group before making the decision on their own.
- *Consultative (CI)* the problem is shared with members of the group on an individual basis, but the leader makes the decision.
- *Consultative (CII)* the problem is shared within the group, but the decision is made by the leader.
- *Group (GII)* the problem is shared by the group, of which the leader is a member (possibly chair), and the decision is made as a result of agreement.

They provide three helpful questions on organisational performance:

1. Do any of the quality requirements make it obviously more rational to choose one solution above all others?
2. What information exists, and is it of a high enough quality?
3. How structured is the problem?

As well as four tests which apply to acceptance:

1. Is it essential for acceptance to occur in order to ensure effective implementation?
2. Will followers accept a decision made only by the leader?
3. How compatible are individual goals with organisational goals?
4. If there is a preferred decision, will it cause conflict in the group?

As Stahl (1995:318) believes, the need to involve all employees in decision making and problem solving is essential in TQ organisations, and he suggests that a leader – participation model will help. But, like others, he also believes that it can seem overly complicated to practising managers and workers. The decision tree has the basis of mathematical inputs (probabilities) and can be computerised. If organisations believe it will assist leaders to make better decisions, leading to improvements, then the model could be extremely useful.

9.4   DIFFERENT APPROACHES TO LEADERSHIP

The sort of leadership that is appropriate when an organisation attempts to implement TQM is a difficult choice to make. Some would suggest that the need for managers who provide leadership in the 'traditional' sense is outdated (Robbins 1994:508). He bases this belief on the fact that organisations which implement TQM, using 'cohesive work groups' (teamworking), will no longer look to the 'great leader' for inspiration and direction. A truly empowered workforce will decide for themselves. However, he does suggest that 'effective leaders don't use any single style' (ibid.:510). Thus leadership becomes a matter of managing change. The transition towards TQM usually involves radical change for any organisation. As Pettigrew and Whipp state:

> Leading change involves linking action by all people at all levels of the business. The primary conditioning features are critical. Early and bold actions can be counter-productive. More promising is the construction of a climate for change while at the same time laying out new directions, but prior to precise action being taken. (1993:6)

Clarke and Pratt (1985) advise that, in order to achieve change, several styles will be required at different stages:

1. *Champion* fights for changes to occur and defends them from critics.
2. *Tank commander* develops strong supportive teams.
3. *Housekeeper* ensures that, as the change process becomes accepted and the culture changes, there is control over essential aspects of the organisation, e.g. cost control and planning.
4. *Lemon squeezer* keeps the workforce motivated and under enough pressure to ensure the organisation can survive.

Rodrigues (1988) supports this view. He argues that organisations experiencing change will go through three different stages. He therefore advises that a different leader with the 'right' skills and abilities is required to deal with the following stages:

1. *The problem-solving stage* requires an innovator, someone who searches for new ideas and who has the 'boldness' to convince others of their appropriateness.
2. *The implementation stage* requires someone who can control and influence people to accomplish the solution provided in stage 1.
3. *The stable stage* is where people need to feel they are able to cope with the new changes. Someone, a pacifier, is needed to ensure that social interaction occurs, and that decentralised decision making is encouraged.

There are others theorists who provide advice for leaders in change situations. Their theories are based upon the belief that if the change to the organisation is going to be radical, then the leadership style should also be radical. Two approaches are usually considered, charismatic and transformational. Both are germane to attempting the sort of cultural change implied by TQM.

9.5   CHARISMATIC LEADERSHIP

Charismatic leadership is about recognising how attribution occurs. Attributes are the characteristics (traits) that people attach to someone. Thus if someone is described as being charismatic, they are believed to possess certain characteristics which make them that way. Crucially, they have traits which mark them out as being special. Politicians like Martin Luther King, John F. Kennedy, Margaret Thatcher and Arthur Scargill have all been described as charismatic. But charismatic people also exist within management. Examples of charismatic business leaders are John Harvey Jones, Richard Branson and Richard Semler. All have been able to create very successful organisations.

The importance of charismatic leaders in organisations is that 'their personal vision, energy, and values inspire followers and thus have a major impact on organizational success' (Stahl 1995:321). But what does being charismatic really mean, and is it possible for any manager to develop the necessary traits?

Robert House (1977), who developed the path–goal theory, has identified three characteristics which make a leader charismatic:

- Extremely high confidence
- Dominance
- Strong convictions in their beliefs

It is worth noting how these traits could well have been applied to Robert Maxwell. Maxwell is now regarded as a corporate bully, even a thief (having plundered the Mirror Group Pension Fund). Charisma is therefore something that can be used for ill as well as good. Thankfully the likes of Maxwell are exceptions. However, we are still interested in how someone can influence others to become committed to actions which will cause the organisation to radically change. The consequence will be that TQM is implemented. Warren Bennis (1984), who studied ninety of the most successful and effective leaders in the United States, found four common features that such managers possessed:

1. They had a compelling vision and a sense of purpose.
2. They could communicate their vision in terms which followers could easily understand and identify with.
3. They displayed behaviour that was consistent and provided an example to followers.
4. They knew their strengths and used them to their fullest potential.

One of the most comprehensive studies into charisma was carried out by Jay Conger and Rabindra Kanungo (1988a). Their findings show that charismatic leaders exhibit the following characteristics:

- Self-confidence
- Vision
- Ability to articulate vision
- Strong convictions about their vision
- Extraordinary behaviour
- Apparent ability to invoke change
- Sensitivity to their environment

Extraordinary behaviour means doing things that appear novel, or unlike the actions of 'normal' leaders.

Being charismatic is clearly desirable if you are leading a change initiative such as TQM. Research by House, Woycke and Fodor (1988) has indicated a correlation between charismatic leadership and satisfaction by followers, leading to improved organisational performance. But is it possible for any manager to learn the art of being charismatic? Anyone who has seen Tom Peters in action will find him a hard act to follow!

Researchers have investigated whether people can learn to be charismatic (Conger and Kanungo 1988b; Howell and Frost 1989). Their findings do indeed indicate that people who have the desire can be trained to:

- Articulate better
- Communicate more forcibly
- Exhibit confidence both in themselves and their followers
- Use direct eye contact and animated facial expressions

There can be a downside to using charismatic leadership for introducing change initiatives such as TQM. According to Machan (1989), if any change was necessitated by a crisis, and the change has solved the problem, charismatic leaders may find their influence is no longer needed. Once things settle down, normal procedures and routines are reintroduced. Advocates would argue that, by its very nature, TQM is continuous (and never-ending), so this should not cause a problem:

> Competition will [also] be improving. . . . Organizations claiming that they have achieved TQM will be overtaken by the competition . . . consequently, the process of quality improvement needs to continuous. This is why TQM should always be referred to as a process not a programme. (Dale and Cooper 1992:14)

9.6 TRANSFORMATIONAL LEADERSHIP

There is some confusion about this approach, because a transformational leader is also charismatic. But transformational leaders go beyond being simply charismatic and avoid the problem identified by Machan.

Bernie Bass (1990) distinguished transactional leaders and transformational leaders. Transactional leaders tend to guide their followers to achieve predetermined goals by clarifying the task and how it should best be accomplished. They are regarded as effective if their followers meet these targets. Normally, no more is expected of them. Most of the theories in this book apply to transactional leadership.

Transformational leaders produce performance in their followers that goes beyond meeting expectations. As Bounds, Dobbins and Fowler describe: 'A transformational leader gets employees excited about the organization's goals and gets them to consider new ways of accomplishing these goals. . . . [Furthermore,] transformational leadership causes employees to become truly committed to the goals of the organization – and take them on as their own' (1995:414).

This means the transformational leader is capable of producing action in others that transactional leaders probably could not. Research indicates that transformational leaders can produce more impressive results far beyond other types of leader (Hater and Bass 1988; Bass and Avolio 1990). According to Avolio and Bass, the difference between a purely charismatic leader and a transformational leader is that the charismatic leader will want followers to adopt their view, whereas 'the transformational leader will attempt to instil in followers the ability to question not only established views but eventually those established by the leader' (1985:14).

This would appear to make the transformational leader eminently suitable for the task of managing a radical change such as TQM. There are many examples in the very successful organisations where the influence of a transformational leader has proved crucial. One such person is David Kearns, who took over as chief executive at Xerox in 1982. Xerox was suddenly discovering that Japanese competitors could sell copiers cheaper than Xerox could actually manufacture them. Kearns realised that drastic action was required in order to survive. He saw quality as being the way forward and instituted a TQM process. Part of this was to benchmark Xerox against Japanese suppliers to learn where improvement could occur (Chapter 11). One critical finding was that Xerox produced products with 30 000 defective parts per million, compared to Japanese equivalents of 1000 per million (Kearns and Nadler 1992; Walker 1992).

Kearns regarded this as intolerable and looked to the workforce for radical improvement. This was achieved through a move away from the traditional command and control towards more cross-functional and participative working in teams. After some major changes, Xerox have been able to compete with Japanese competitors on both price and quality. Indeed Xerox UK won the prestigious European Quality Award in 1992 (Chapter 12).

SUMMARY

The issue of leadership is crucial to the success of quality management. Managers in organisations who wish to ensure that initiatives are likely to be successful must lead from the front and by example. It is therefore incumbent on senior and middle management to attempt four things:

1. Avoid the use of traditional command and control.

2. Act as facilitators to encourage the efforts of workers.

3. Be willing to listen to suggestions, no matter where they come from.

4. Try to emulate successful leaders (transformational leaders).

QUESTIONS

Discussion

Consider an impressive leader you have come across recently. Think of the main aspect that was most inspiring about this person. Using a single piece of paper shared by the whole group, put down all the aspects you have thought of.

(a) How many characteristics are there?

(b) Are some more important, and why?

(c) Is it possible for one person to possess all of them?

Individual

The chairperson of an apparently underperforming company has approached you to get advice on how their organisation may be transformed. You find that all the managers have been employed on the basis of coming from the right schools. Write a report of up to six pages covering the following items:

(a) Provide specific guidance on how to change the way in which the organisation's employees are led?

(b) What would you tell the chairperson to do about the criteria for selecting leaders (managers) in the future?

FURTHER READING

Beckhard, R. and W. Pritchard (1992) *Changing the essence: the art of creating and leading fundamental change in organizations*. Jossey-Bass, San Francisco CA.

Bounds, G. M., G. H. Dobbins and O. S. Fowler (1995) *Management: a total quality perspective*. South-Western College Publishing, Cincinnati OH.

Kakabadse, A., R. Ludlow and S. Vinnicombe (1988) *Working in organisations*. Penguin, London.

Katzenbach, J. (1996) *Real change leaders*. Nicholas Brealey, London.

Langford, D., M. R. Hancock, R. Fellows and A. W. Gale (1995) *Human resources management in construction*. Longman, Harlow.

Lavender, S. (1996) *Management for the construction industry*. Addison Wesley Longman, Harlow.

Taffinder, P. (1995) *The new leaders: achieving corporate transformation through dynamic leadership*. Coopers and Lybrand/Kogan Page, London.

Tichy, N. M. and M. A. Devanna (1986) *The transformational leader*. John Wiley, New York.

CHAPTER 10

USING QUALITY IMPROVEMENT TECHNIQUES

OBJECTIVES

- Describe the improvement techniques used in conjunction with quality management initiatives.
- Differentiate between hard and soft approaches to quality management.
- Describe the hard aspects of quality management and how they can be applied.
- Explain the importance of teamworking in TQM.
- Describe models for encouraging teamwork.

10.1 CONTEXT

Using quality improvement techniques in order to introduce TQM involves what Wilkinson (1994:274) calls 'hard' and 'soft' sides. The hard side includes the use of tools which have definite applications. The soft side involves using tools in a way that emphasises the people in an organisation. All advocates of TQM agree that teamwork is the best way to achieve success.

10.2 TOOLS FOR TQM

These tools are used in order to collect data which will help to understand problems, find their causes and assist in developing solutions. There are a variety of tools. The purpose of each tool depends upon the context and requirement for a particular solution. As Oakland explains (1993:216), problems are opportunities to 'focus on improvement [which] leads to the creation of teams whose membership is determined by their work and detailed knowledge of the process.' He provides a strategy which uses a flowchart (Figure 10.1).

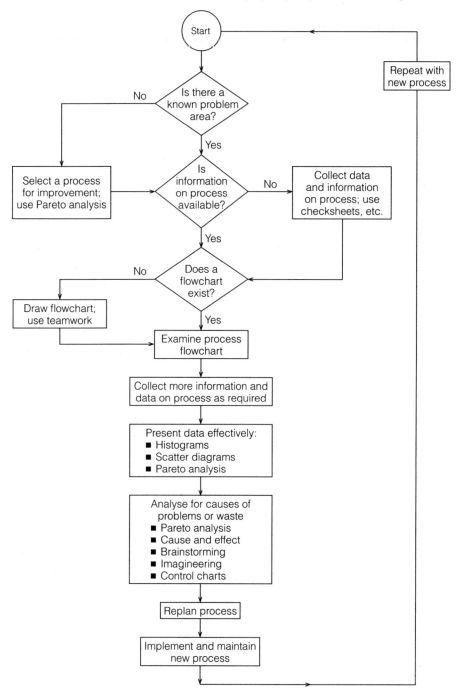

Fig. 10.1   A strategy for process improvement (Reprinted with the permission of Butterworth-Heinemann, from Oakland 1993:217)

10.2.1   Checklists

Checklists are a simple means of providing an *aide-mémoire*. It tells the user there are certain things which must be checked. As such, it can be used in the auditing of QA and to follow the steps in a particular process.

10.2.2   Flowcharts

Flowcharts, also called process flowcharting, give a diagrammatic representation of the stages or steps in a process. They provide a picture of the tasks that are being performed. Because they are diagrammatic, flowcharts are usually much easier to understand than a written description. Those who actually carry out the tasks are usually best placed to spot problems and areas where improvement is possible.

Dale, Bunney and Shaw (1994:383) list the steps for constructing a flowchart:

1. Define the process, its boundaries, and the start and finish.
2. Decide the method of charting and the symbols to be used, then stick to them all the way through.
3. Agree the amount of detail to be used.
4. Give brief descriptions of each stage.
5. Get those who carry out the process to check the flowchart for accuracy and logicality.

10.2.3   Checksheets

Checksheets differ from checklists in that they are used to record events, or non-events (non-conformances). They can also include information such as the position where the event occurred and any known causes. They are usually prepared in advance to be completed by those who are carrying out the operations or monitoring their progress. The value of checksheets is they enable retrospective analysis, so they help with problem identification and problem solving.

10.2.4   Tally charts and histograms

Tally charts and histograms are used to compile information on the frequency and patterns of variation for tasks where variability is expected. The aim is to measure the variation so it can be minimised. Tally charts and histograms should be well designed so they can be easily used by people who carry out the operations.

10.2.5   Scatter diagrams

Scatter digrams plot the relationship between two variables. Often easy to complete, an obvious linear pattern reveals a strong correlation.

10.2.6   Pareto analysis

Pareto analysis attempts to prioritise problems so that attention is initially focussed onto those which have the greatest effect. It is actually named after a nineteenth-century Italian economist, Vilfredo Pareto, who observed how the vast majority of the wealth (80%) was owned by relatively few of the population (20%). This has been adopted as a generalised rule for considering solutions to problems, i.e. that 80% of problems emanate from 20% of the causes.

Pareto analysis aims to identify the 'critical' 20% of causes and to solve them as a priority. The belief is that most of problems will then be eliminated. Data is collected on checksheets about problems and their causes. Then a cumulative frequency curve is plotted to identify which causes generate the greatest numbers of problems.

10.2.7 Cause and effect analysis

Also called a fishbone diagram, cause and effect analysis was originally developed by Karou Ishikawa to break down the major causes of a particular problem. The end result looks like the skeleton of a fish. This is because a process often has a multitude of tasks feeding into it, any one of which may be a cause. If a problem occurs, it will have an effect on the process, so it will be necessary to consider the whole multitude of tasks when searching for a solution.

Cause and effect diagrams (Figure 10.2) tend to be used in conjunction with other techniques such as brainstorming and quality circles. Certain steps should be followed when preparing a cause and effect diagram:

1. Gather a team of workers who contribute to all the tasks which make up the process.
2. Ensure they all understand the nature of the problem.
3. Define the symptoms of the problem, i.e. its effects.
4. Consider the potential causes.

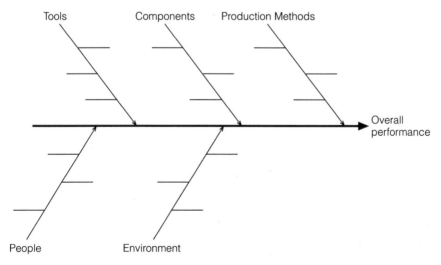

Fig. 10.2 Cause and effect diagram: also called an Ishikawa diagram or a fishbone diagram (Thomas 1995:73)

5. Draw up the cause and effect diagram.
6. Use a brainstorming session to speculate on the reason for these causes, and in turn the way in which they can be solved (ranking may be useful to prioritise).
7. As a team, formulate plans of action to implement solutions in accordance with the system of ranking.

The cause and effect diagram has been further developed by Fukada at the Sumitomo Electric plant. This derivation is known as the CEDAC approach because of the addition of cards (cause and effect diagrams with addition of cards). The cards provide a means by which the team can update their facts and ideas about the problem. The aim is to quantify the causes of the problem, hence to measure reduction due to improvement efforts.

10.2.8 Brainstorming

Brainstorming is an approach to problem solving based on the principle that free expression is a necessary prerequisite to getting adventurous and creative ideas. Although it is extremely beneficial to cause and effect diagrams, it can be applied to any TQM technique.

The philosophy is that any person who puts forward an idea, free from ridicule or criticism, will spark ideas in the rest of the group. In order to achieve an atmosphere conducive to creating solutions, all those who are brainstorming must be regarded as being equal (status outside the group is suspended). There are some basic rules:

1. The team should be clear about the nature of the problem.
2. The team should appoint a facilitator.
3. There should also be someone who will record ideas. A flipchart can be useful because sheets can be hung around the room to provide a reminder of ideas.
4. There should be rules of conduct:

   - Each member to offer one idea at a time, in rotation.
   - Keep ideas short and simple, they are easier to record.
   - Stimulate for innovative and free-flowing ideas.
   - Seek appropriate clarification but offer no judgement of suggestions.
   - Keep the atmosphere informal, the process should be fun.

5. Be aware of problems:

   - Attempting to get straight to solutions
   - Criticism of outlandish ideas
   - Those who want to take over or impose their ideas
   - Destructive arguments
   - Unproductive discussion or several competing discussions

The importance of brainstorming is that it can generate many ideas. These ideas can be evaluated, analysed and ranked. The most popular solutions will be those which have greatest consensus. Implementation is therefore something that the group have agreed, and should certainly be less problematic than solutions imposed by those on high.

10.2.9 Nominal group technique

The nominal group technique is a form of brainstorming where the aim is to maximise the number of people, the number of levels and the number of teams. It is particularly useful where communication across suborganisational boundaries could be a problem.

A facilitator begins by reading out a statement of the problem to be solved. The groups can then ask questions to clarify the problem and restate it in their own words. All groups then follow this sequence:

1. *Silent idea generation*: each person thinks about how they might solve the problem.
2. *Idea collection*: each person briefly recounts their thoughts.
3. *Solution consideration*: some contributions may need to be explained more fully and some contributions may be discarded.
4. *Individual ranking*: the facilitator asks each person to rank the remaining solutions on scorecards.
5. *Overall ranking*: the facilitator collects the cards and produces an overall ranking.
6. *Solution adoption*: the highest-scoring solution is adopted; further discussion will be required in the event of a tie.

10.2.10 Quality function deployment (QFD)

Quality function deployment (QFD) is a technique which originated in 1972 in Japan, at the Kobe shipyard of Mitsubishi. It is essentially a method by which a product or service is designed to suit customer expectations in every aspect of its technical requirements. The objective is to have a continuous stream of information fed into the process of development from concept through to actual use, and back again. QFD thus incorporates:

- Market research
- Basic research
- Invention
- Concept design
- Prototype testing
- Final product or service testing
- Aftersales service and troubleshooting

Any organisation using QFD will attempt to anticipate the future needs of their customers. It is therefore not surprising that QFD is so closely associated with the Japanese, who have demonstrated what paying attention to detail and customer satisfaction can achieve.

QFD uses the House of Quality, so called because it looks like a house (Figure 10.3). Thus the house is a table which brings together the answers to three questions about customers: Who are they? What are their needs? How will their needs be satisfied?.

The whats are recorded in rows, and the hows in columns. The product or service is considered as a whole before examining its constituent parts. This technique is standard for any motor car, and all of its components. A House of Quality is constructed in seven steps:

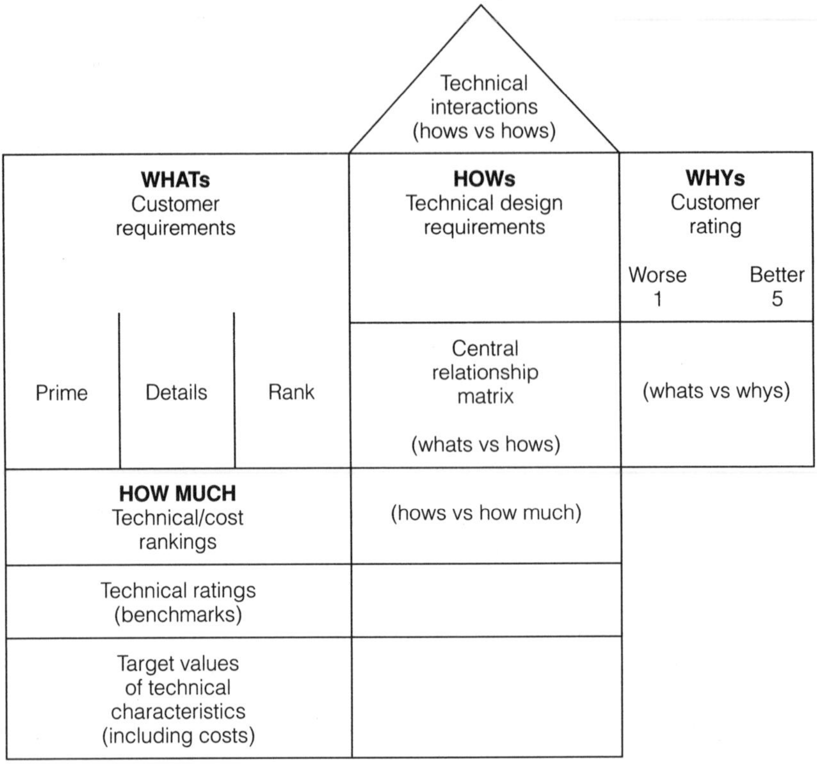

WHATs Customer requirements			HOWs Technical design requirements	WHYs Customer rating
				Worse 1 Better 5
Prime	Details	Rank	Central relationship matrix (whats vs hows)	(whats vs whys)
HOW MUCH Technical/cost rankings			(hows vs how much)	
Technical ratings (benchmarks)				
Target values of technical characteristics (including costs)				

Technical interactions (hows vs hows)

Fig. 10.3 The House of Quality (Reprinted with the permission of Butterworth-Heinemann, from Oakland 1993:50)

1. Customer requirements (whats) are obtained using marketing and research, i.e. focus groups; attempt to consider radical features which may create excitement and give pleasure.
2. The customer requirements are classified into small numbers of primary, secondary and tertiary whats.
3. The whats are now ranked to establish their importance by the customer. The results are entered into the box to the right of the central matrix.
4. The whats must now be translated into hows, the technical design requirements. The hows are then ranked in terms of cost and rated against any whats the competitors provide. The whats will usually require competitive benchmarking (Chapter 11). This information will allow the QFD team to evaluate the efficiency and cost of the hows that are put forward.
5. The core to the diagram, the central relationship matrix, is where the whats are compared with the hows. It is where each customer requirement is considered against the technical design requirement. Symbols are used to indicate the strength of the relationships between them (usually strong positive, neutral, negative and strong negative). This will indicate the potential for achieving the whats using the hows.

6. The roof shows the interactions between the hows. The objective is to ensure they are compatible, so the end product is what the customer will want.
7. The box containing the target values of technical characteristics is what each solution will need and an estimate of its cost.

QFD is a continuous process which is likely to bring about the following benefits:

- Improving overall quality
- Increasing customer satisfaction
- Cutting down the design time and rework
- Lowering overall costs
- Improving reliability
- Improving communication

It is important to stress that QFD can be applied to any application, not just manufacturing. It can also be used in the strategic planning of a business. The objective will be to set goals for the divisions and in turn the departments.

10.2.11   Failure mode, effect and criticality analysis (FMECA)

FMECA aims to determine the possibility of failure in products or services, and to predict the effect. The objective is to ensure that, once any potential failures have been identified, corrective action is taken to reduce and eradicate the possibility of reoccurrence. Although FMECA can be applied at any stage, it is most logical and cost-effective to carry it out at the design stage. Its main elements are as follows:

- *The mode of the failure* will indicate when and where it is most likely to occur; this could be in a particular component or at an interface between parts. Alternatively, there may be particular conditions which increase the likelihood of failure.
- *The effect of the failure* is a prediction of what will happen if a failure occurs.
- *The failure criticality* assesses the degree of severity should a failure occur.

In order to use FMECA, certain items need to be judged on a scale of 1 to 10:

- P is the probability of failure (1 = low, 10 = high)
- S is the seriousness of the failure (1 = not serious, 10 = very serious)
- D is the difficulty of failure detection before the product or service is actually used by the end customer (1 = very easy, 10 = very difficult)

From this information it is possible to calculate a risk priority number (RPN):

$$RPN = P \times S \times D$$

This will indicate the priority of the failure; failures with the highest priority need the most attention.

10.2.12   Moment of truth (MOT)

The moment of truth (MOT) is a concept associated with failure. It was created by Jan Carlzon of Scandinavian Airways. Essentially, an MOT is the moment a customer first comes into contact with an organisation's people, systems, products, methods, and so on.

There is a widespread belief that first impressions count. Thus, if a customer has a poor MOT, it is going to take a lot to redeem the organisation, its products or its services in that customer's eyes. Consequently, MOT analysis aims to reduce any potential failures which might occur at the moment of truth. This will require the proactive involvement of the people who actually interface with the customers.

10.2.13   The seven 'new' tools of quality

Seven tools have been developed by the Japanese to collect and analyse problems that are often regarded as difficult to quantify:

1. *The affinity diagram* is used to assemble the problems which may cluster around a particular area; it is produced by the following steps:

   - Decide an area or theme.
   - Collect all relevant data from customers, personnel involved, suppliers, etc., and identify the problems which are believed to exist. Record the ideas on cards or Post-it notes.
   - Display the ideas at a meeting to see whether anyone likes them.
   - Get those involved to select the ideas which they regard as being related to different sub-areas or sub-themes; this will provide ideas for solution by other tools.

2. *The interrelationship diagram* is often used after the affinity diagram. The objective will be to take the ideas and map out logical and sequential links. It allows those involved to plot what they think happens, even if they're not sure. It is often regarded as a free-flowing cause and effect diagram.

3. *The systems flow/ tree diagram* produces a logical chronology of all the functions that make up an operation or task.

4. *The matrix diagram method* indicates the relationship between objectives and methods, and results and causes of the problems identified. It is similar to the central objective of the House of Quality. It aims to evaluate the relative importance of problems.

5. *Matrix data analysis* is where information gathered is plotted and analysed to show clearly what problems or areas need to be tackled first. It is very useful in marketing and product research.

6. *The process decision programme chart (PDPC)* aims to select the most appropriate processes which will contribute to the desired outcome. Thus, if a problem is identified, all the possible causes and the potential outcomes are evaluated. The objective is to think about as many unexpected problems as possible, and to plan countermeasures for them.

7. *The arrow diagram method* should be familiar to any student of planning techniques. Usually called the critical path method or the precedence diagram

method, it is the way to break down a project, i.e. a product or service, and to plot the interrelationship between tasks. As a result, it becomes possible to evaluate the most critical activities – to identify those activities which will have most impact on the overall project. It is also possible to consider the implications of changing resources for each activity.

10.2.14   Statistical process control (SPC)

Statistical process control (SPC) is perhaps one of the most significant quality tools, due to its development at the Bell Laboratories by Shewhart in the 1920s. It was subsequently used by Dr Deming as a central part of his philosophy, which the Japanese adopted as part of their drive towards quality.

Probably because it uses statistics, SPC seems to appear off-putting to all but those who regularly handle data. In fact, in its simplest form, SPC can be used for any process. The objective is to measure the inputs against the output. Even the most trivial of administrative tasks can be treated as a process. The important thing is to collect data on the parts that can be measured, e.g. the time taken from start to finish, or the number of defectives in a process. Using simple formulas it is then possible to establish the control limits, upper and lower, on what is known as the control chart (Figure 10.4).

What is most important is that variation in the process can then be assigned to either special or common causes. As Figure 10.4 shows, if the variation is within the control limits, it is due to common causes; outside the limits it is due to special causes. Perhaps you remember from the description of Deming's philosophy in Chapter 4, special causes indicate something exceptional. Providing they are aware of the problem, those most directly involved with a process will usually be best placed, to rectify any variation due to special causes.

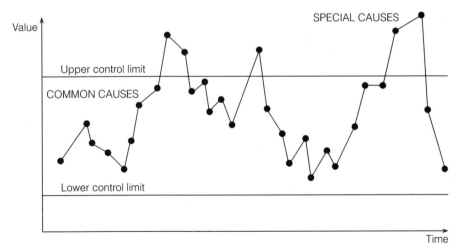

Fig. 10.4   The SPC control chart (Reprinted with the permission of Prentice Hall Europe, adapted from Logothetis 1992:231)

Common causes are those which are beyond the influence of workers. They are due to the system in which they work. Crucially, if there is to be any improvement, it is only the management who can do something. As Deming came to believe, common causes can be of the order of 98%.

Those who compile the chart can use it to analyse whether the process is stable, to determine what variation is occurring, and most crucially, to see who is responsible for remedying any problems. It is thus possible to identify what needs to be done to create improvement.

If the process is within the control limits, it is said to be under control. By applying SPC to implement continuous process improvement, the Japanese demonstrated how control limits can be narrowed, significantly reducing the vast majority of variation, which usually causes the problems. In fact, the ultimate goal is to sustain a process without any variation. As Deming advocated, this will require the workers and management to cooperate. The aim will be to develop the process to continually adapt to changing customer needs, while ensuring the output is as free from variation as possible.

10.2.15 Quality costing

All of the tools which have been described so far attempt to give an indication of the size of problems, and where improvement is possible; the cost of quality is just one more item to be quantified. In particular, so-called non-quality costs arise when problems could have been avoided in the first place. Some estimates of their magnitude place them at one-third of an organisations' expenditure (Dale and Plunkett 1990). This indicates a massive reduction in potential profit, or alternatively a huge competitive market advantage if the savings are passed onto customers. As Dale states, 'The importance of definitions to the collection, analysis and use of quality costs cannot be overstressed' (1994c:210). He goes on to explain some of the many reasons for collecting quality-related costs (ibid.:213–14):

- To display their importance to the organisation in a meaningful way.
- To indicate their relative importance on key business areas.
- To assist in identifying areas for improvement, especially non-value-adding activities.
- To enable comparisons to be made.
- To allow budgetary control.

The method by which these costs are collected is entirely up to the organisation, but help can be found in BS 6143 *Guide to the economics of quality-process cost model*. Dale and Plunkett's *Quality costing* (1991) provides some excellent advice. The most useful information is not always immediately obvious. Indeed, it may be well hidden. People are very good at covering up things which they fear may be used against them. Consequently, any effort to identify costs, or to set up a quality costing system, should aim to encourage active and honest participation.

The objective is to consider the overall process, not just the end result. The amount of defectives is an obvious starting-point, but it is only a starting-point. Many of the tools described up to now would be perfectly effective for establishing quality costs. As so often pointed out, try to seek contributions from those most directly involved in the processes. They have vital knowledge about how the processes work and where there is room for improvement, perhaps leading to cost savings.

10.3 TEAMWORK FOR TQM

According to Oakland, teamwork is vital because it 'changes the independence to interdependence through improved communications, trust and the free exchange of ideas, knowledge, data and information' (1993:319). Figure 10.5 shows the advantages of teamwork.

In modern organisations the processes undertaken are usually so complex and involve so many different departments or individuals that teamworking is not only desirable, it is essential. According to Dale and Boaden (1994:515), there are many good reasons for using teams:

■ Coming together reinforces commitment to TQM
■ Beneficial to communication
■ More consensual decision making
■ Fostering of trust and cooperation
■ Problem sharing to reach effective solutions
■ Building collective responsibility
■ Promoting cutlural change for TQM

Dale and Boaden also point out that teamworking is widely regarded as the reason for the success of Japanese suppliers. They also draw attention to three types of team: project, quality circles and improvement teams.

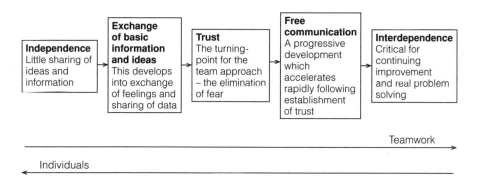

Fig. 10.5 Teamwork takes you from independence to interdependence (Reprinted with the permission of Butterworth-Heinemann, adapted from Oakland 1993:319)

10.3.1   Project teams

Project teams usually have a temporary objective. They also tend to be led by someone with seniority. The completion of the task is often the only real aim, all other issues being secondary. Consequently, this type of team (very popular in construction) is not one that is really regarded as being able to contribute to improvement. The only way to do so would be to have improvement as the most important objective. The project leader must also be prepared to be more of a facilitator than a line manager.

10.3.2   Quality circles

A quality circle is usually a voluntary group of people who come together to solve problems as a team. A successful quality circle needs to have the following characteristics:

1. Contribution is encouraged, not forced.
2. Everyone has equal status.
3. Although there is no ideal number, six to eight is desirable; nine or more will probably be too big.
4. Items for discussion or problems to be solved are selected by the group.
5. Meetings are short but frequent, perhaps weekly and lasting no more than one hour.
6. A facilitator is available at the outset.
7. The group is capable of implementing its own solution.

The group will also need to possess the following skills (training by the facilitator may be necessary):

1. Data collection and analysis using the tools described earlier in the chapter.
2. Administration of the meeting and subsequent documentation.
3. Presentation of findings in either report form or direct to senior management.

10.3.3   Improvement teams

Improvement teams are different from quality circles because their objective is to consider improvement in general. Likely topics include productivity, motivation, cost and waste reduction. The main aim of these deliberations is to ensure that whatever is implemented will add value for the customer.

An improvement team is likely to have some of the following characteristics:

1. Membership may be voluntary or mandatory.
2. Subjects for discussion may be decided by management.
3. The team leader may be appointed by the team or by management.
4. The improvement team is usually more temporary than a quality circle.

Table 10.1   Quality circles versus yield improvement teams

Feature	Quality circles	Yield improvement teams
Purpose	Involve employees Increase employee participation in the business Team building Develop people	Improve process yields Reduce scrap Solve quality-related problems
Team building	Will only solve problems if an effective team has been developed Members work together Operate by consensus	Formed around a problem Members are given specific tasks Onus is on the individual Peer pressure to perform Goals, targets and achievements are established and assessed Team develops around its achievements
Leadership	Section members/first-line supervisor Members lack authority and power Lack of access to functions, people and information Dependent on others for data and advice	Production/section managers Members are relatively senior Independent
Problem-solving potential	Limited Minor problems Limited skills	Considerable Major problems Highly skilled
Project resolution rate	Low	High
Infrastructure	Steering committee Infrequent meetings Lack of regular reporting of individual circle progress	Steering committee Monthly reports to managing director Weekly reporting of leaders to steering committee

Source: Reprinted with the permission of Prentice Hall Europe, from Dale and Boaden (1994:522).

Manson and Dale (1989) have tabulated the main differences between quality circles and 'yield' improvement teams, a type of quality improvement team (Table 10.1). The chosen teamwork solution should be suited to the objective. Quality circles and yield improvement teams have the following characteristics in common:

1. *A team sponsor* helps to get it going, resolves problems or conflicts, and coordinates the activities of the team so they are compatible with the activities of the organisation.

2. *A team facilitator* acts as a coach and enables members to actively contribute. They may also be responsible for the team's ability to communicate with other areas and parts of the organisation.
3. *A team leader* is responsible for organising the meetings and their agendas. They will also ensure that meetings run smoothly, i.e. according to agenda and to time, and that follow-up actions are minuted for later verification.

10.4 SETTING UP A TEAM

Getting the right team is important. The ability of the team to produce results will bear testimony to how well the team was set up. But what are the characteristics of an effective team, and how can the most suitable team be assembled?

Dale and Boaden (1994:525) provide the answer to the first part of the question:

1. Every member is an active and willing participant.
2. The relationships are good.
3. There is trust and respect among members.
4. All ideas are welcome.
5. There is commitment to the working/principles of the team.

They also provide some useful guidelines to assist in building a successful team:

1. Those who will be affected by the results of the team are included in the team activity.
2. All members are aware of and educated about quality management, particularly the use of improvement tools.
3. Every person is clear about the aims and objectives of the team.
4. There is discipline and adherence to agreed rules.
5. All members should regularly attend.
6. Periodic reports to senior management should be encouraged (this will recognise the worth of the team).
7. There is regular evaluation during and after the team's life to review what feedback can be gleaned. This will be useful information in other teams.

10.5 STAGES IN TEAM DEVELOPMENT

Tuckman and Jensen (1977:419–27) suggest there are four stages in the development of a team:

1. *Forming* aims to make everyone aware of what is required; getting people to conform and understand each other.
2. *Storming* is the stage where conflict is likely; this conflict has to be dealt with in order to make any progress.
3. *Norming* is where the group will start to gel, and cooperation will occur; the team will start to display confidence, trust in each other, clarification about their

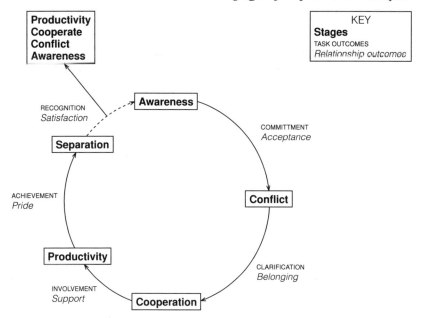

Fig. 10.6 Team stages and outcomes (Oakland 1993:336)

purpose, an ability to think about problem solving, and consideration of how improvement can be achieved.

4. *Performing* occurs when the group is actually producing.

These stages are summarised in Figure 10.6, which Kormonski and Mozenter (1987) developed as a modification to the work of Tuckman and Jensen.

10.6 ROLES WITHIN A TEAM

It is unusual for every member of the team to do the same thing. The team requires different people to take on different tasks. This belief was developed by Dr Meredith Belbin as a result of his work in industry and in management training. In his 1981 book, *Management teams: why they succeed or fail*, he explains there are eight roles which will give the team a greater chance of success:

1. *Coordinator*. This is essentially a chairperson. It requires a strong person who can impose discipline on the team, but it should not be someone who wants to impose their own will.
2. *Shaper*. This person will provide the stimulus to the group's efforts. They will listen to the discussions and shape them into practical solutions.
3. *Plant*. This person provides the original and creative ideas upon which solutions can be built.

4. *Monitor/evaluator*. This person will analyse the solution being discussed and can predict the practical implications. They will stop the team committing itself to solutions which are unworkable or unfeasible.
5. *Implementor*. This is the person who looks at the team's objectives and puts them into a logical order. In effect, they have the capability to turn words into action.
6. *Resource investigator*. This person will be able to ascertain what resources are available to the group. They act as a scout, having links with other departments and teams.
7. *Teamworker*. This person is the hub of the group. They will want to operate within a cohesive and coherent unit because they see that strength comes from unity. There may be several teamworkers in one group.
8. *Finisher*. This person will keep the pressure on to see that any ideas or solutions come to fruition. They will not allow the group to walk away just because the team feels it can fire and forget.

Belbin makes some important additional recommendations:

1. All team members should be equal; no one should be perceived as a star or be given greater importance.
2. One person can take on different roles in a team.
3. Inward-looking roles are

   ■ Coordinator
   ■ Plant
   ■ Resource investigator
   ■ Shaper

4. Outward-looking roles are

   ■ Implementor
   ■ Monitor/evaluator
   ■ Teamworker
   ■ Finisher

Belbin provides a questionnaire that will help to determine the role which best suits each team member. But it may also be worth considering the use of MBTI, described in the next section.

10.6.1   The Myers–Briggs type indicator (MBTI)

The MBTI is an aid to developing teams (Briggs 1987). It is based on four scales which analyse an individual's preference for

■ Giving and receiving 'energy'
■ Gathering information
■ Making decisions
■ Handling the outside world

Each scale is based on two opposites:

- *E*xtroversion versus *I*ntroversion
- *S*ensing versus i*N*tuition
- *T*hinking versus *F*eeling
- *J*udgement versus *P*erception

Thus a person completing a questionnaire (which must be administered by a qualified MBTI facilitator) will arrive at one of sixteen combinations of four letters from the four scales of preference. This will help to establish which role best suits that team member.

10.6.2 Oakland's DRIVE model for quality through teamwork

DRIVE is a structured method for problem solving. Oakland (1993:349) describes the title as a mnemonic, because the letters come from the following steps the team must carry out to 'keep on track and in the right direction':

- *D*efine the problem, so the team is agreed on what is to be done.
- *R*eview the information available.
- *I*nvestigate how the problem can be solved and what means or resources are required.
- *V*erify the proposed solution.
- *E*xecute the chosen solution.

SUMMARY

Tools and techniques are tangible aspects of quality management. Without them, any improvement initiative will be based on hope rather than sound data. It is the combination of the so-called hard tools, in conjunction with the soft methods of developing teamwork which will determine success. The following points should be remembered:

1. Problem solving should become a prime objective of any improvement effort.
2. An appropriate tool should be used to collect, analyse and develop solutions to problems which obstruct quality.
3. The techniques should be simple but effective, and all employees should be trained in their application, especially workers at the operational level.
4. Any improvement tool or technique should be used within a teamwork situation; this will encourage collective effort and consensus.

QUESTIONS

Discussion

The acronym TEAM spells out the phrase, together each achieves more.
Discuss the perception that individualism is frequently regarded as a virtue which

is encouraged in all aspects of life. Are there any situations when teamworking will be inappropriate within an organisation? Why?

Individual

This requires an applied solution and ideally will mean gaining access to a real work situation.

Having gained permission to use either a department within your organisation, or your educational institution, identify a task that is carried out on a regular basis. Use a flowchart to analyse what is achieved and how. Then collect some data on how effectively the objectives have been achieved. Using some simple statistics, provide a report on how you believe the task may be made more efficient. If it is possible, talk to those who actually carry out the task; better still, involve them in writing your report.

(a) What are your findings?

(b) Were you told about the previous efficiency of the task before you began your investigation? If so, what was it?

(c) If you have been able to implement and monitor a solution, how much improvement did you discover?

(d) If you have involved the work group, what were their feelings about the way in which they had been able to influence their work in the past? Had they been encouraged to work in teams or to use problem-solving tools?

FURTHER READING

Hutchins, D. (1985) *Quality circles handbook*. Pitman, London.

Jay, R. (1995) *Build a great team*. Pitman, London.

Kolarik, W. J. (1995) *Creating quality: concepts, systems, strategies, and tools*. McGraw-Hill, New York.

Lewis, L. (1984) *Quality improvement handbook*. McGraw-Hill, New York.

Lock, D. (ed.) (1990) *Gower handbook of quality management*. Gower, Aldershot.

Mizuno, S. (1988) *Management for quality improvement: the seven new QC tools*. Productivity Press, Cambridge MA.

Ozeki, K. and T. Asaka (1990) *Handbook of quality tools*. Productivity Press, Cambridge MA.

White, A. (1996) *Continuous quality improvement: a hands-on guide to setting up and sustaining a cost-effective quality programme*. BCA Books, London.

Wickens, P. (1987) *The road to Nissan: flexibility, quality, teamwork*. Macmillan, London.

CHAPTER 11

OTHER WAYS TO ACHIEVE IMPROVEMENT

OBJECTIVES

- Consider techniques to support improvement efforts.
- Describe how partnering can be developed with clients, suppliers and subcontractors.
- Explain the use of benchmarking to identify areas for potential improvement.
- Understand how business process re-engineering can radically reorganise core processes to make them more efficient.

11.1 CONTEXT

There are other techniques which can be used by organisations to attempt quality improvement. They are complementary to TQM, so they can be regarded as ways of embedding any initiatives. It is not crucial to try them all simultaneously; there is the danger of quality fatigue! However, some can be seen as helping to facilitate long-term and sustainable cultural change.

The previous chapter described tools and techniques of achieving improvement. These tend to be internally focused. It is also necessary to consider aspects of improvement which are externally focused. In particular, it is essential to collaborate with bodies that are normally outside the control of the organisation attempting to make improvement. This is because most organisations rely on others to contribute materials, components, services and often labour. Accordingly, any effort towards improvement will be limited by the exclusion of these suppliers. It is therefore prudent to consider how they can assist any improvement efforts. The methods described in this chapter may be of assistance. Three areas are worth considering:

- Partnering
- Benchmarking
- Business process re-engineering (BPR)

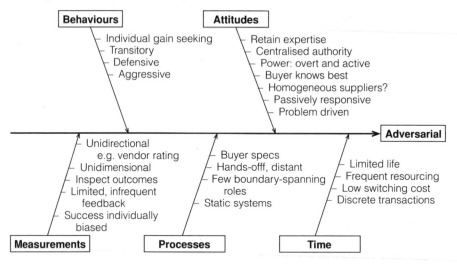

Fig. 11.1   Adversarial relationships (Macbeth and Ferguson 1994:124)

11.2   PARTNERING

11.2.1   What is partnering?

'An important aspect of the overall change brought about by a TQM approach is a changed relationship with suppliers' (Sadler 1995:102). Sadler explains that the 'traditional' approach of being involved in 'antagonism' leads to conflict-ridden relationships. This has seldom produced quality. At best, he believes, it reduces cost. However, more effort is put into arguing over contractual details rather than making sure the end result is the best that can be achieved. Sadler calls for 'sustained collaborative relationships with suppliers.' This is the concept of partnering, also called 'co-makership' (Philips 1985).

Using cause and effect diagrams (Figures 11.1 and 11.2), Macbeth and Ferguson (1994) contrast the principles of 'adversarial' and 'collaborative' relationships: There are many well-known examples of superlative partnerships in action. One is the relationship that Marks & Spencer (M&S) has with its suppliers. M&S has been described as a 'manufacturer without factories' (Scarbrough and Corbett 1992:154). Even though M&S do not actually control the supplying organisations, they are able to influence them to innovate and create new products. The merchandise that M&S retails is known for being of high quality. Suppliers have to assure M&S that their products will be consistent. Both sides enjoy an obvious benefit. If the products sell well, M&S increases its profit and raises its reputation; this has occurred consistently in the past. The suppliers benefit by selling more of their products. This also helps to establish a long-term relationship for developing new products.

Car manufacturers have learned the lesson that long-term improvement will be more likely if suppliers of the basic components are involved. It should come as no

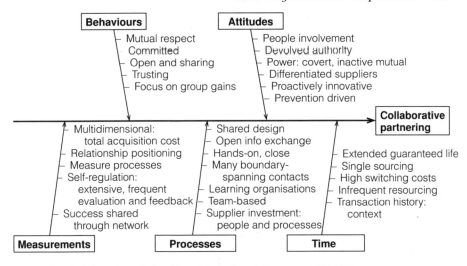

Fig. 11.2 Collaborative relationships (Macbeth and Ferguson 1994:124)

surprise that it is the Japanese who have led the way in this regard. In their influential book *The machine that changed the world*, Womack, Jones and Roos describe the way in which Japanese manufacturers have used lean production techniques to create relationships based on 'mutual benefit, rather than . . . mutual suspicion' (1990:148). In another revealing book, *The Japanisation of British industry*, Oliver and Wilkinson state their belief that: 'Japanese manufacturing methods dramatically increase the interdependencies between actors involved in the whole production process, and that these heightened dependencies demand a whole set of supporting conditions if they are to be managed successfully' (1992:68).

11.2.2 What are the features of partnering?

Dale, Lascelles and Lloyd provide a description of what they call 'supply chain management' (another name for partnering). They define this concept as a situation of 'working together towards a common goal' (1994:294). It leads, they believe, to 'benefit through cooperation, rather than pursuing self-interests; a win–win situation' (ibid.:294). They list the typical features:

1. Long-term contracts based on

   ■ Common aims, particularly to continuously improve the product or service
   ■ A shared desire to work together
   ■ Mutual trust
   ■ Cooperation
   ■ Honesty
   ■ Open declaration of problems

2. A willingness between both parties to learn more about each other. This may involve members of supplier and client organisations working at the other's offices or sites.
3. A constant interchange of information, including all financial matters.
4. Cooperation on the design and methods of production at the earliest opportunity.
5. Joint problem solving to achieve best practice.
6. A reduction in the size of the supplier base

11.2.3   How to develop partnering

Macbeth and Ferguson suggest there are five phases to what they call 'relationship improvement' (1994:164); they are illustrated in Figure 11.3.

Phase 1: internal commitment and team building

This consists of bringing about a change within the customer organisation. As Macbeth and Ferguson believe, it is often one person, a 'champion' (the quality manager) who will do this. It will be necessary to address the following issues:

- Strategies to develop partnerships
- The benefits and implications of developing long-term relationships
- Change in the attitude of internal managers towards suppliers

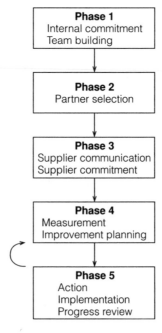

Fig. 11.3 The five phases of relationship improvement (Adapted from Macbeth and Ferguson 1994:164)

Phase 2: partner selection

Partnering arrangements usually work better with fewer suppliers. Otherwise adequate communication and a relationship built on trust cannot occur. Dale, Lascelles and Lloyd (1994:309) give the probable benefits of reducing the number of suppliers:

- Less variation in product characteristics
- More time for QA needs of vendor (buyer)
- Simpler and better communication with vendors
- Less paperwork
- Less transportation
- Less handling and inspection

Without a reduction in suppliers, paperwork, transportation and inspection will all increase, leading to higher transaction costs (Macbeth and Ferguson 1994:97). Cutting down on suppliers is a matter of examining the abilities of existing organisations. By considering the following sort of questions it may be possible to 'produce a short list of . . . around 10 per cent of the original number' (ibid.:97):

1. Are partners strategies compatible for forming long-term collaborative relationships?
2. Is the partner willing to reorganise to meet customer needs that are going to be necessary in the future?
3. Has the partner the sort of expertise that will be capable of integrating?
4. Does the partner have access to innovative technology, or markets which will assist in the long term?
5. Can the partners systems and procedures be integrated?
6. Is there a basis for the partnership to work on a personal basis?

Phase 3: supplier commitment and joint team building

By the time this stage is reached, it is likely that many aspects of the partnership will have been discussed informally. Although partnerships are founded on trust, some sort of formal commitment is beneficial to demonstrate goodwill and intent. This can be by a partnering agreement, which according to Macbeth and Ferguson (1994:171), includes the following items:

- Statement of principle
- Time of partnership
- Likely or potential targets
- Agreement to share gains and losses
- Expectations of each partner
- Confidentiality
- Review procedures
- Mutual termination

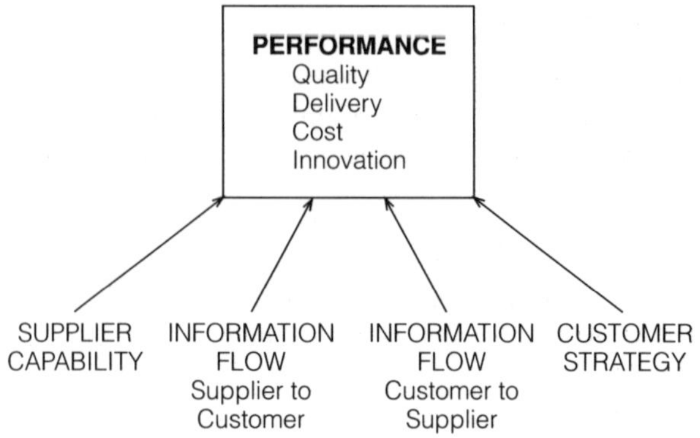

Fig. 11.4   The relationship positioning tool (Adapted from Macbeth and Ferguson 1994:173)

Phase 4: measurement and improvement planning

This is where the relationship comes into its own. The aim will be to get benefits at the earliest opportunity. It is important to ensure there is a basis for measuring the effects of what is being achieved. This will then allow modification to capture best practice and to ensure that future improvement is possible. Figure 11.4 shows the 'relationship positioning tool' which Macbeth and Ferguson (1994:173) liken to a tree. The 'leaves on the tree (*performance measures*) . . . indicate overall health' of the relationship. But these leaves depend on the 'strength of the root system.' Each of the root headings will cover a variety of issues.

Customer strategy

- *Requirements*: quality, delivery time/pattern, costs the market will accept, innovation.
- *Attitude towards partners*: business relationship, long-term commitment, degree of involvement, interdependency.
- *Systems to support*: appropriate costing, evaluation of output, approval of output, scheduling, involvement on a day-to-day basis.
- *People*: organisational structure, involvement in terms of training and education (a basic requirement of TQM), ways of recognising effort.

Supplier capability

- *Profile and strategy*: customer bases, organisation, plans for improvement, financial health, flexibility and recognised expertise in its output.
- *People that partners employ*: skills, involvement in decision making and improvement, training, flexibility and competence, overall attitude to committed relationships.
- *Process techniques and systems to meet customer expectations*: design, capability, capacity, achievement in a way which adheres to principles of safety and quality.
- *Supplier management aspects*: supply base, selection and approval of its own suppliers, evaluation of output, communication, involvement with others, delivery costs and quality.

Information flow   Information flow is two-way and in any partnership it is essential that everyone is committed to the exchange of information. The following issues will need to be covered:

- *Technical*: customer specifications, capabilities, problems, improvements and innovations; adherence to quality is paramount.
- *Involvement*: teamwork, feedback from each side, dedication to sharing.
- *Business relationship issues*: cost sharing, lead-in times, open-book management (sharing of financial information).
- *People issues*: absolute support and commitment to make the partnership work, appropriate organisational structure and interpersonal links (don't expect people who cannot get along to work effectively).

Phase 5: action implementation and progress review

This is a retrospective consideration of how the relationship has performed in phase 4. It is crucial to realise some achievement, or progress towards it. Failure to achieve, without good reason, will lead to serious questions about the future viability of the relationship. Here are some likely issues:

- Achievement of anticipated targets
- Agreed methods for measurement
- Procedures for reporting
- Communication mechanisms
- Ways of highlighting success
- Review of continued relationship and setting of new targets
- Increased socialisation of employees from partner organisations

Increased socialisation means arranging sports events and family outings. The objective is that people see the relationship as beneficial beyond the simple desire for commercial success. Many Japanese organisations believe this to be important.

11.2.4   Potential problems with partnering

Any relationship such as partnering does not happen overnight. It will take much time and effort to establish. Similar to the development of TQM, it may be years rather than months before any significant benefits occur. However, dangers are always possible in any relationship requiring 'mutual dependency and high levels of trust' (Macbeth and Ferguson 1994:189). The main pitfall is that one side sees themselves as being wronged. The result will be a collapse of all the effort, particularly trust and goodwill.

All partners should be aware of this problem and seek to avoid it. Macbeth and Ferguson suggest that other external threats may occur (1994:190):

1. A new competitor emerges providing a cheaper product or service. Good benchmarking and intelligence will help to avoid this situation.

2. The market changes very quickly and new expertise or technology is required. The temptation may be to change partners to deal with it. Once again, it is part of the partnership relationship to try and stay ahead of the game; benchmarking and intelligence are vital.
3. End customers demand significant price reductions. This may seem an inevitable part of modern business. The lesson from the Japanese and their followers is that long-term relationships, especially the sharing of expertise, can both increase quality and reduce costs.

Finally there may be internal threats. Internal threats are more likely to stem from misunderstandings, conflict between employees belonging to different partners, and a sudden change due to losing a vital member of the interorganisational team. There is no foolproof way to avoid any of them. The main objective is to set up a relationship that ensures everyone sees there is more to gain from collaboration than getting embroiled in disputes.

11.3   BENCHMARKING

Sadler believes it is possible for an organisation which has instituted an improvement initiative such as TQM to 'settle into a state of complacency combined with arrogance' (1995:84). This he thinks is likely if the only comparison it makes, is with past performance rather than 'the best in the world'.

An organisation needs to make sure it is keeping up with its competitors, ideally by comparing itself against 'proved performers', a process known as benchmarking. The benchmarking organisations may come from any business sector.

Benchmarking, according to Oakland, is believed to have originated in Japan and is based on the ancient Japanese quotation: 'If you know your enemy and know yourself, you need not fear the result of a hundred battles' (1993:180–81). Oakland goes on to define the word *dantotsu* as 'striving for the best'; *dantotsu* is the Japanese for benchmarking.

Sadler explains that benchmarking is a useful part of any TQM process. It aims to identify 'best practice' in other organisations, particularly those regarded as the best, so that 'objective comparisons' can be made with what you do (1995:85). As he stresses, 'It is when the gaps between such best practice and [your] own standards are shown to be alarmingly wide that a climate for change is created.' The purpose is to establish the critical success factors, crucial aspects of the business for altering competitiveness. Figure 11.5 illustrates the benchmarking cycle.

The result of identifying such critical success factors is to ensure you can apply them to your own organisation. Your desire is to use these critical success factors so they will increase your ability to provide superior quality to end customers. This is that main reason for benchmarking. Table 11.1 sets out the likely objectives of an organisation attempting to improve. It also shows the difference between using benchmarking and not using benchmarking. There are three main types of benchmarking:

Fig. 11.5 The benchmarking cycle (Reprinted with the permission of Prentice Hall Europe, from Bank 1992:33)

Table 11.1 Reasons for benchmarking

Objectives	Without benchmarking	With benchmarking
Becoming competitive	Internally focused Evolutionary change	Understanding of competition Ideas from proven practices
Industry best practices	Few solutions Frantic catch-up activity	Many options Superior performance
Defining customer requirements	Based on history or gut feeling Perception	Market reality Objective evaluation
Establishing effective goals and objectives	Lacking external focus Reactive	Credible, unarguable Proactive
Developing true measures of productivity	Pursuing pet projects Strengths and weaknesses not understood Route of least resistance	Solving real problems Understanding outputs Based on industry best practices

Source: Reprinted with the permission of Butterworth Heinemann, from Oakland (1993:181).

- Internal
- Competitive
- Functional or generic

11.3.1 Internal benchmarking

As its name implies, internal benchmarking compares all the operations within an organisation. It can be achieved in the early stages of TQM when one department, office or site has made extremely good progress. Other parts of the organisation seek to emulate this progress, and will visit the location to see what has been done, and how. Internal benchmarking seldom has problems with access, a distinct advantage.

11.3.2 Competitive benchmarking

This involves making a comparison with those organisations that operate in the same area or sector. As a result, they are likely to be competitors. It is also called reverse engineering or teardown. This is because in the manufacturing industry, if your competitors come up with a new, or radically improved product, the most obvious way to investigate is to buy one and pull it apart. This will allow you to see what components are used, and how. However, this probably won't reveal much about the processes they used. Some inside knowledge will be needed!

Cross and Leonard warn there is a danger of 'more of the same and a me-too strategy' (1994:502). Straightforward 'copying' is unlikely to achieve radical transformation in the perceived quality of an organisation's products or service.

11.3.3 Functional or generic benchmarking

This is regarded as the purest form of benchmarking. It involves the comparison of a function, or indeed the whole organisation, against another organisation regarded as superior in its field. The objective is not to copy but to be inspired.

Non-competitor organisations are more likely to release information on their processes. Cross and Leonard reinforce this view, 'It is encouraging how open a business will be if asked to share its success stories' (1994:503). This can easily be verified by consulting the literature on TQM and benchmarking. Surprisingly many organisations have provided revealing information on how they were able to achieve radical transformation. This was traditionally held as a commercial secret, guarded at all costs. This new-found freedom of access indicates that many organisations believe they're a long way ahead and they are likely to remain at the front because, in the spirit of *Kaizen*, their improvement effort will be continuous.

11.3.4 The Rank Xerox experience

Probably the most widely cited example of benchmarking is Rank Xerox. Earlier I described David Kearns and his influence as a transformational leader at Rank Xerox. He set in motion the Leadership through Quality strategy in 1983, a year after his appointment as chief executive. Rank Xerox had until then enjoyed a virtual monopoly in the market for photocopiers. However, Japanese electronics companies saw this as a market in which they could compete. A study of likely Japanese competitors led Xerox to realise that drastic action would be required. An early example of competitive benchmarking, the Xerox study revealed a need for an immediate response. Xerox had discovered that 'the Japanese were selling their machines for what it cost Xerox to make them. . . . [Up until the report, Xerox had assumed] the competitors' machines were poor quality. This was proved by benchmarking to be wrong and to drive the point home they [the Japanese] were making profit!' (Cross and Leonard 1994:498).

These findings led the senior management at Rank Xerox to acknowledge the magnitude of the threat they were facing. Fortuitously, a Japanese subsidiary, Fuji

Xerox, was performing significantly better; they were reaping the benefits of the Deming philosophy. Rank Xerox therefore had the advantage of being able to carry out internal benchmarking against their Japanese subsidiary. This led to its TQM initiative, initially something of a survival plan. But TQM did more than that; it enabled Xerox to become a producer of high quality competitive products. This was confirmed in 1992 when the UK subsidiary won the highly regarded European Quality Award.

Besides its TQM initiative, Rank Xerox also carried out a generic benchmarking study on American mail-order company L.L. Beens, who despite having 'an entirely different product mix and operat[ing] in a different industry, [provided] the logic of order processing, stockkeeping and invoicing routines' (Karlof and Ostblom 1993:48). Beens were a 'best practice' company and provided many valuable lessons on how Rank Xerox could radically improve its ability to satisfy customers.

11.3.5   Using the Rank Xerox benchmarking process

Rank Xerox has produced a ten-step, five-phase model (Figure 11.6) for benchmarking in a 'disciplined and structured way' (Camp 1989). Cross and Leonard advise that, in using this model, 'while each phase is critical and needs to be completed thoroughly, the more time that is devoted to the planning stage the less time is likely to be wasted later on' (1994:503). It is also a 'process [which] is dynamic and requires some flexibility in application.' But they stress how the objective is to complete the process in a commonsensical way. This will provide the organisation with the critical success factors that will enable it to compete more effectively in the future. Each of the five phases contains some essential steps.

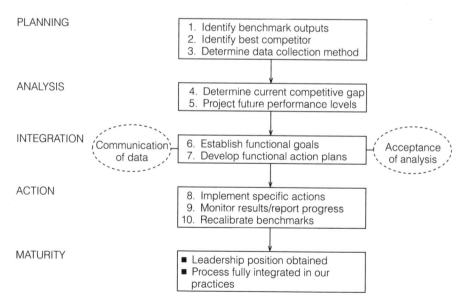

Fig. 11.6   The Rank Xerox benchmarking process (Adapted from Cross and Leonard 1994:504)

Phase 1: planning

Planning is crucial. The decisions taken here will generate the data upon which the future actions are based. Several questions should be raised:

1. What areas are to be benchmarked?
2. Which organisations can provide potential benchmarks?
3. Are there possible data collection or methodology problems?
4. Who will be involved?
5. Is there a limited amount of time?

Cross and Leonard provide some useful guidance: 'Benchmarking is viewed by Rank Xerox as an active learning experience, the value of [which] comes from seeing and understanding how [others] achieve their performance levels' (1994:505).

They also advise that just carrying out benchmarking will not 'automatically result in actions to improve.' They believe it is essential to have a 'clear and systematic rationale beforehand.' It is important to be realistic about what it is possible to achieve and 'a prerequisite to benchmarking [is] to really understand in detail one's own processes.'

Certain areas of the organisation may offer lessons on improvement which can usefully be learned. However, if it is necessary to go to other organisations, be sure to find out what can or cannot be compared. Going in blind will produce ineffective benchmarking. This is known euphemistically as industrial tourism.

Phase 2: analysis

Having carried out the benchmarking exercise, certain things should start to become clear: how much better are the other organisations; the areas of most concern; what can be done to make improvements in the short, medium and long term.

The analysis should be made on 'good quality data' (the result of intelligent planning). It is important that everyone realises the truth. If the benchmarking reveals deep problems, they should be explicitly confronted there and then. Alternatively, if the exercise confirms a situation of relative comfort, in terms of competitive advantage, the staff should be commended. A favourable outcome should also act as the spur for continued improvement. Complacency is a sure way to lose any advantage.

Phase 3: integration

The findings of the benchmarking will need to be put into effect. The chosen method is very important, particularly in terms of communication. It may be that some decisions are sensitive or very radical. Care is required to enable those directly affected to say how they believe any outcome could best be achieved. Consensus and support are without doubt the keystones of any improvement initiative.

Phase 4: action

Actual implementation of the findings is the moment when improvement begins in earnest. It will require those involved to be fully informed, committed and capable. Staff and operatives must possess requisite skill and expertise, and be given all necessary resources. Detail plans are important and key personnel should be identified. The plans should be flexible enough to allow appropriate alterations. Benchmarking, like TQM, is a continuous process.

Phase 5: maturity

Maturity arrives when the benchmarking process has become 'an integral part of the management process' (Cross and Leonard 1994:510). Benchmarking no longer needs to be initiated by senior managers or as the result of perceived crisis. All employees regard it as an essential response to changing markets and increased competition. Karlof and Ostblom suggest that benchmarking can exert a force that engenders a cultural change, a force which 'shifts the focus of attention of people and organizations from all kinds of trivialities to what is fundamental to their individual and collective success' (1993:180).

The sort of environment they recommend is where benchmarking is accompanied by benchlearning; constant learning and relearning become the norm. This is summarised in Figure 11.7. Notice how it refers to the learning organisation. This concept is advocated by Senge (1990) and Argyris (1993) and will be described in chapter 12.

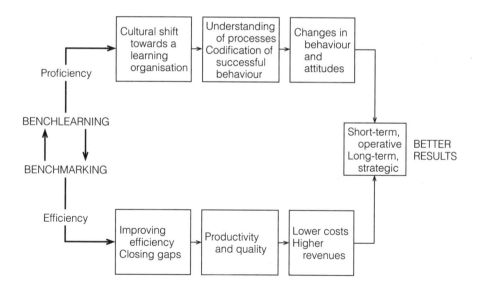

Fig. 11.7 Benchmarking and benchlearning as simultaneous activities (Adapted from Karlof and Ostblom 1993:181)

11.4 BUSINESS PROCESS RE-ENGINEERING

Sadler defines business process re-engineering (BPR) as 'an approach to radical or transformational change which focuses on questioning the need for and the means of carrying out each of the many processes involved in the organization's tasks' (1995:111). At the heart of the concept is the focus on processes. Processes are significant because, in any organisation, there are several 'steps' which have to be carried out, often one after another. The tradition has been that each step completes a specialised task. The output of each step is then sent to the next step. The focus in each part of the overall process is 'functional hierarchy' (Peppard and Rowland 1995:7). The result is that the people employed in each step (department) will normally only be interested in aspects that are relevant to their part of the process.

Accepting that various tasks may continue to exist, BPR emphasises the need to 'cut across the functional hierarchies to reach the customer' (ibid.:7). BPR is used to ensure that 'the customer's interests come first rather than what is most convenient for particular organizational sub-units' (Sadler 1995:111). The objective is thus to improve the effectiveness of an organisation in 'meeting customer needs while eliminating waste of materials, capital, and the time of people' (Bell, McBride and Wilson 1994:56). As they believe, BPR is capable of 'improving competitive position while enhancing process adaptability to ensure continuing relevance to changing business needs.' There are five characteristics of a high quality process:

1. It meets customer requirements effectively, and any other criteria.
2. It uses resources efficiently.
3. It is under control; it behaves predictably with minimal variability.
4. It can be monitored to detect changes.
5. It adds value.

The last point is important because, unless it can be demonstrated that a process is adding some value, it is pointless to perform it. Many organisations experience the problem that nobody has ever taken an overall view. In particular, there is often a reluctance to ask, Why is that done like this? But this is the way that it has always been done, is no justification.

The important aspect of BPR is that people who carry out tasks or operations are encouraged to look beyond their own functions. As Bell, McBride and Wilson suggest, managing processes and managing people are very similar (Figure 11.8). Coulson-Thomas (1996:25–26) lists the main principles of BPR, many of them compatible with the ideas behind TQM:

1. Focus on increasing the amount of value added throughout the process.
2. Give the customer a single and accessible point of contact.
3. Internally focus on harnessing the potential of people and applying it to those activities which identify and deliver value to customers.
4. Encourage learning and innovation in a creative environment.
5. Use project teams for horizontal management which concentrates on flows.

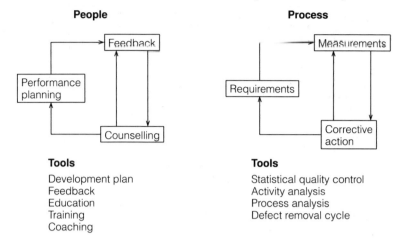

Tools

Development plan
Feedback
Education
Training
Coaching

Tools

Statistical quality control
Activity analysis
Process analysis
Defect removal cycle

Fig. 11.8 The similarities between managing people and managing processes (Reprinted with the permission of Butterworth-Heinemann, from Bell, McBride and Wilson 1994:57)

6. Remove any non-value-adding activity.
7. Link performance measures and rewards to customer-related outputs.
8. Redefine the manager's role so that emphasis on command and control gives way to empowerment and the notions of coach and facilitator.
9. Follow through item 8 by giving discretion and authority to those people closest to the customer.
10. Encourage participation and involvement.
11. See simplification as a virtue and minimise the number of core processes.
12. Build learning, renewal and short feedback loops into processes.
13. Ensure that continuous improvement is built into implemented solutions. BPR can reawaken interest in TQM; the two complement each other.

11.4.1 How to carry out BPR

Peppard and Rowland (1995:204) propose five key phases for BPR (Figure 11.9):

1. Create the environment.
2. Analyse, diagnose and redesign the processes.
3. Restructure the organisation.
4. Pilot and roll-out.
5. Realise the strategy.

Create the right environment

This involves getting everyone to accept the need for change. It is likely to be something that many will not appreciate. People will experience various stages

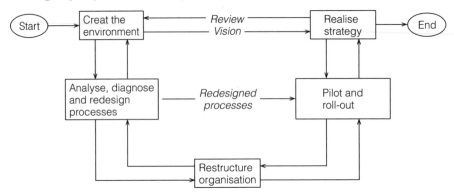

Fig. 11.9   An overall approach to BPR (Reprinted with the permission of Prentice Hall Europe, adapted from Peppard and Rowland 1995:204)

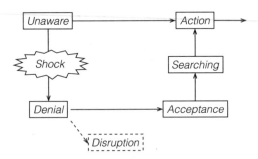

Fig. 11.10   General reactions to change (Reprinted with the permission of Prentice Hall Europe, from Peppard and Rowland 1995:206)

in any change process such as BPR; these stages are normal and they are shown in Figure 11.10.

It will take time for most people to go through all the stages. When the majority have accepted the need to change, only then will real progress be possible. A small number may refuse. This will be unfortunate, but their ability to disrupt is something that must be carefully considered. If the whole initiative runs the risk of being undermined, there needs to be a decision on how much longer they can be tolerated.

Much will depend on the ability of managers to provide both vision and enthusiasm for change. As Peppard and Rowland (1995:212) recommend, this places several requirements on senior management:

1. Build a BPR vision of an improved organisation adding extra value.
2. Exhibit total commitment.
3. Formulate and communicate a clear plan.
4. Provide appropriate training and education.
5. Identify the core processes.
6. Appoint a champion who will instil enthusiasm and elicit support.

Analyse, diagnose and redesign the processes

The objective is to get those who actually operate the processes to understand what is being done and why. Having gained this understanding it is then possible to get front-line people to carry out the tasks of re-engineering the key processes. Harrington (1991) suggests that tools will make this easier, tools such as flowcharting, interviewing of task operatives, elimination of non-value-adding and bureaucratic activities. He particularly advocates using simple language.

Peppard and Rowland (1995:216) provide nine steps to carry out this phase:

1. Recruit and train teams.
2. Identify process outcomes and linkages.
3. Analyse existing processes and quantify measures.
4. Diagnose conditions for improvement.
5. Benchmark best practices from elsewhere.
6. Redesign processes in order to optimise.
7. Review people's contribution to 'new' processes.
8. Review the technological inputs.
9. Validate the new processes to ensure they actually provide added value.

Restructure the organisation

When the technologies have changed and the people have changed, the organisational structure will probably need to change. Peppard and Rowland (1995:220) recommend six steps:

1. Review the organisation's people resource in terms of structure, competences and motivation.
2. Review the structure and technological capabilities.
3. Formulate a new organisational form which is 'appropriate'.
4. Define new roles for all those involved.
5. Deal carefully with those whose jobs may no longer exist because certain processes are no longer needed.
6. Build new technological infrastructure and applications.

The use of IT is regarded as essential in all organisations. BPR will give an opportunity to install new systems to provide a communication infrastructure. In the fourth item, Peppard and Rowland stress that managers should behave like coaches, training but not directing.

Pilot and roll-out

The new processes are tested by those who must operate them in the new structure. The best way of doing this is to do it for real. By selecting customers, the BPR initiative can be given its launch. The benefit of not going all-out at this stage lies in being able to identify potential problem areas, perhaps to iron them out. As Peppard and Rowland explain, 'It is during this pilot period that organizations will seek to come down the learning curve of the new process quickly so that lessons can be incorporated into other areas before the complete overhaul of the re-engineered organization is completed' (1995:222).

They provide the following steps for this phase:

1. Select the processes to be piloted and match them to the appropriate customers.
2. Put together the pilot team who will monitor and report on progress.
3. Ensure that lessons from piloting are fed back quickly into other processes.
4. Prioritise the actual implementation (roll-out) once these refinements or amendments have been incorporated.

Realise the strategy

This is when the re-engineered organisation actually operates the new processes and structures as a matter of routine. The test of success will be the way in which those people who work for the organisation approach their tasks following the changes. If employees are genuinely supportive and see benefit, it will increase the long-term potential for giving added value to customers. The nature of any change process is always going to depend on people. Ignoring their input is bound to cause problems. Peppard and Rowland have adapted advice from V. E. Schein (1985:39–40) to assist with BPR:

1. Make the initiative non-threatening.
2. Be truthful about why the BPR initiative is required and explain its potential benefits and problems.
3. Engage in open discussion with those who are opposed; don't stifle opposition.
4. Recognise that it is the middle mangers who will actually put it into operation. Their understanding, support and belief is crucial.
5. Bargain and trade-off because the long term is more important than getting everything in place at once, with the potential for hostility if change is imposed.
6. Treat the initiative as an experiment to reduce the perception that senior mangers have all the answers.
7. Begin small and expand the initiative on the basis of success rather than attempting a Big Bang.

Finally, Coulson-Thomas has this to say about BPR, 'The best form of long-term insurance against failure or irrelevance, and protection against errors of judgment or perception, is to build learning loops into processes and create a learning community' (1996:85). He uses a diagram to demonstrate the connection between organisational vision, capability and customer requirements (Figure 11.11). In particular, he believes there must be 'processes for continuous learning and improvement.'

As with benchmarking, there is justification for learning in any organisation that implements BPR. The creation of 'learning organisations' is increasingly advocated as part of developing quality management. The result will be an environment that encourages long-term improvement. This concept is examined in the next chapter.

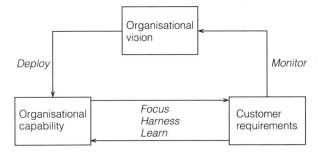

Fig. 11.11 The connections between organisational vision, capability and customer requirements: (focus) processes for focusing on delivery of value to customers; (harness) processes for harnessing talents of groups and teams to add value to customers; (learn) processes for continuous learning and improvement (Reprinted with the permission of Kogan Page Ltd., from Coulson-Thomas 1996:86)

SUMMARY

This chapter has described three methods to support a quality management initiative such as a TQM:

1. *Partnering* seeks to gain a greater understanding of other parties who contribute to the whole process. This will involve clients, which will allow an organisation to know exactly what they want. It also encourages the involvement of suppliers and subcontractors to the organisation from which the end customer consumes.

2. *Benchmarking* aims to compare the way in which key processes are carried out elsewhere (selected because of acknowledged excellence). The aim is to learn how others achieve their tasks so effectively. This will indicate what needs to done to be more competitive and will help to create improvement.

3. *Business process re-engineering* is a method of looking afresh at how an organisation carries out its key operations. The objective is to ensure that all actions and processes achieve quality for customers – they are value adding.

Anything that does not add value should be questioned then avoided in future.

QUESTIONS

Discussion

One of the problems that appears to afflict the construction industry is a reluctance to cooperate with others. Discuss whether or not this is true and consider the relevance of any methods in this chapter. What steps should be taken to encourage all contributors to work together?

Individual

Imagine you are a newly appointed manager in an organisation. You have been told that a benchmarking exercise is considered a good method for analysing competitiveness. Write an essay of about 2000 words covering all three items:

(a) What would you advise, and which other type of organisation should be benchmarked?

(b) Provide a flowchart of the key tasks to be carried out, and indicate what each task would involve.

(c) What sort of time period should be dedicated to the process?

FURTHER READING

Baden-Hellard, R. (1995) *Project partnering: principles and practice*. Thomas Telford, London.

Bendell, T., L. Boulter and J. Kelly (1997) *Benchmarking for competitive advantage*, 2nd edn. Pitman, London.

Champy, J. and M. Hammer (1993) *Reengineering the corporation*. Nicholas Brealey, London.

Coulson-Thomas, C. (1996) *Business process re-engineering: myth and reality*. Kogan Page, London.

Karlof, B. and S. Ostblom (1993) *Benchmarking: a signpost to excellence in quality and productivity*. John Wiley, Chichester.

Lamming, R. (1993) *Beyond partnership: strategies for innovation and lean supply*. Prentice Hall, Hemel Hempstead.

Lascelles, D. M. and B. G. Dale (1993) *The road to quality*. IFS Publications, Bedford. Chapter 6 on supply-chain management is particularly interesting.

Peppard, J. and P. Rowland (1995) *The essence of business process re-engineering*. Prentice Hall, Hemel Hempstead.

Porter, L. and S. Tanner (1996) *Assessing business excellence*. Butterworth-Heinemann, Oxford.

Regan, J. (1995) *Crunch time: how to reengineer your organization*. Century, London.

Zairi, M. (1996) *Benchmarking for best practice: continuous learning through sustainable innovation*. Butterworth-Heinemann, Oxford.

CHAPTER 12

LOOKING TO THE FUTURE: IS THERE LIFE AFTER TQM?

OBJECTIVES

- Develop awareness of techniques and methods that go beyond TQM.
- Understand what is meant by the learning organisation.
- Appreciate how the Toyota principles have evolved into lean management.
- Describe the six levels of TQM adoption.
- Appreciate the use of quality awards in becoming a world class organisation.

12.1 CONTEXT

This Chapter describes aspects of quality management which, although they are not strictly more sophisticated or beyond TQM, can still be regarded as the next step. Specifically, they seek to build upon what will already have been implemented in any TQM initiative. They aim to bring out the full potential of people and their ability to provide superior quality.

In their book, *Beyond total quality management: towards the emerging paradigm*, Bounds et al. (1994:5) describe how managers have found that 'traditional approaches to management are inadequate for keeping up with changes.' As they suggest, increased global competition and improved communications have meant that customers expect more. Managers of organisations have frequently responded by implementing TQM. However, the experience for many has been disappointment. This is because TQM has often been treated as just another programme to provide quick results. Indeed, Harari (1993:32–35) provides ten reasons why TQM has proved to be disappointing for some organisations:

1. It focuses people's attention on internal rather than external results.
2. TQM aims at achieving only minimum standards.
3. It can develop its own cumbersome bureaucracy.
4. TQM delegates quality to tsars and experts rather than real people.
5. It often does not demand radical reform.
6. TQM often seeks only short-term financial gain.

161

7. Relationships with outsiders, i.e. suppliers, are often little different to those which existed before.
8. TQM appeals to faddism. egotism and quick-fixism.
9. TQM drains entrepreneurship and innovation form corporate culture.
10. TQM has no place for love.

Furthermore, suggest Bounds et al., 'Implementing another program with a three-letter acronym is not the way to stay competitive. Rather transformation is required. Traditional managers must think and act differently. The changes required of them are profound. Managers do not need another program, but another paradigm' (1994:6).

They point out the significance of the word *paradigm* as defined by Thomas Kuhn (1962). Commenting on scientific revolutions, Kuhn stated that a paradigm was the 'universally recognized scientific achievement that for a time provide model problems and solutions to a community of practitioners.' Bounds et al. (1994:6) believe that paradigms in management 'consist of the way people think and act in conducting business.' The consequence is that managers must engender a shift in the 'values, beliefs, traditional practices, methods, tools . . . that members of a social group [organisational members] construct to integrate [their] thoughts and actions.'

In the introduction to what is undeniably a long book (over eight hundred pages), Bounds et al. provide the basic principles of the 'emerging paradigm beyond TQM' which managers need to be aware of in order to 'think and act to improve organizational systems to provide superior customer value' (ibid.:5). In order to explain these principles, they describe it in terms of three themes:

- Customer value strategy (Table 12.1)
- Organisational systems (Table 12.2)
- Continuous improvement (Table 12.3)

Table 12.1   Old and new paradigms for the customer value strategy

Topics	Old paradigm	New (emerging) paradigm
Quality	Meeting specifications, inspected into product, make trade-offs among quality, cost, schedule	One component of customer value, managed into process, seek synergies among quality, cost, schedule
Measurement	Internal measures of efficiency, productivity, costs and profitability, not necessarily linked to customers	All measures linked to customer value
Positioning	Competition	Customer segments
Key stakeholder	Stockholder, boss (other stakeholders are pawns)	Customer (other stakeholders are beneficiaries)
Product design	Internal, sell what we can build	External, build what customers need

Source: Bounds et al. (1994:29).

Table 12.2 Old and new paradigms for organisational systems

Topics	Old paradigm	New (emerging) paradigm
Cross-functional approach	Negotiation across functional interfaces to obtain cooperation	Cross-functional systems defined, owned and optimised
Technology	To deal with complexity, to eliminate people problems	To reduce complexity, source of optimisation for customer value
Employee involvement	Focused on hygiene factors	Focused on strategic factors
Human resource management	Regarded as a staff responsibility, administration of personnel hiring, firing and handling complaints	Regarded as a critical resource, managed as system input
Role definition	Task and job descriptions set limits	Vision inspires flexibility
Culture	Social and emotional issues are suppressed, politics and power dominate	Connect with individual sense of purpose, emotions and social meaning
Structure	Specialisation, tall hierarchy with functional emphasis	Integration, flat hierarchy with team emphasis

Source: Bounds et al. (1994:31).

Table 12.3 Old and new paradigms for continuous improvement

Topics	Old paradigm	New (emerging) paradigm
Occasion	Focused new product development, episodic, reactive to problems, big breakthroughs only	Focused on broader systems, unending, proactive to opportunities, big breakthroughs and small steps
Approach	Trial and error	Scientific method
Response to error	Punish, fear, cover-up, seek people fix, employees are responsible	Learning, openness, seek process/ system fix, management is responsible
Decision-making perspective	Individual political expediency, short-term	Strategic, long-term, purposeful for organization
Managerial roles	Administer and maintain status quo, control others	Challenge status quo, prompt strategic improvement
Authority	Top-driven via rules and policies	Customer-driven through vision, enablement and empowerment
Focus	Business results through quotas and targets	Business results through capable systems, means tied to results
Control	Scoring, reporting, evaluating	Statistical study of variation to understand causes
Means	Delegated by managers to staff and subordinates	Owned by managers who lead staff and subordinates

Source: Bounds et al. (1994:31).

In addition Bounds et al. draw attention to many other aspects they regard as essential to underpin the 'emerging paradigm'. This chapter aims to describe three possible ways to look beyond TQM:

- The learning organisation
- Lean production and Management (the Toyota production system)
- Becoming world class by using quality awards and prizes

12.2   THE LEARNING ORGANISATION

This is a concept that has become increasingly popular as a means to stimulate radical transformation. As Burnes (1996:191) believes, learning is necessary as a prior condition to change, specifically the organisation must encourage every member to 'become familiar with the marketplace, customers, competitors, legal requirements . . . in order to recognise the pressures for change.' I have alluded to this theme many times already. However, the learning organisation is one which Pedler, Burgoyne and Boydell (1991:1), believe 'facilitates the learning of all its members *and* continually transforms itself.'

Sadler explains how this implies that all activities carried out are 'monitored to provide feedback which is then used to improve performance' (ibid.:123) Pettigrew and Whipp (1993:18) suggest the most important part of organisational change is 'collective learning'; failure to understand this, they believe, will 'render [any improvement initiative] virtually useless.' They go on to suggest how 'the use of such learning ensures that the full implications of the firm's view of its environment are captured, understood and retained at all levels.'

The significance of organisational learning is that managers must encourage a questioning attitude, such that every action or event is analysed to see whether there is a better way in the future. Sadler (1995:124) believes that learning takes place on three levels:

- Individuals experience events and reflect upon how and why they occurred.
- Teams are where groups of individuals learn things together.
- Organisations learn through the cumulative experiences of their individuals and teams; organisational learning aims to systematise best practice.

As Sadler also describes, the transfer of learning within an organisation can be of two types:

- *Direct (immediate)* transfer of knowledge is by in-company training, counselling, coaching and appraisal, etc.
- *Indirect (deferred)* transfer of knowledge occurs when knowledge is accumulated and stored in things like procedure manuals and systems. The so-called culture of the organisation is the result of this accumulation.

The diagram which accompanies Sadler's definitions (Figure 12.1) shows the processes and influences of organisational learning.

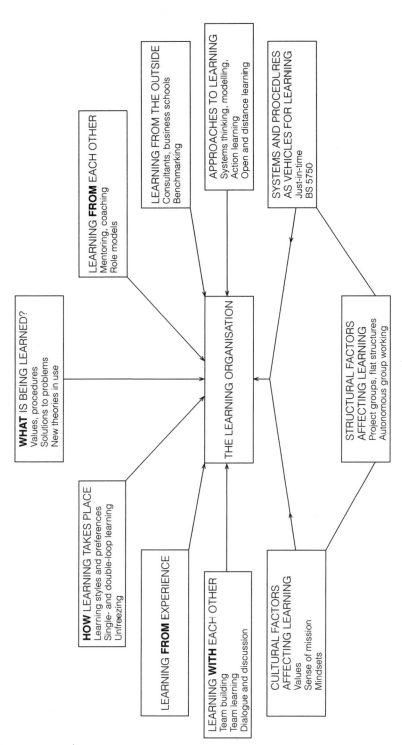

Fig. 12.1 Learning organisations: processes and influences (Reprinted with the permission of Kogan Page Ltd., from Sadler 1995:126)

The most widely recognised exponent of the learning organisation is Peter Senge. This is a result of his book, *The fifth discipline: the art and practice of the learning organization*, published in 1990. In it he stressed his belief that the organisations which are most likely to succeed in the future will be those which utilise people's capacity to learn at every level. Accordingly, managers must be willing to communicate openly and frankly. The emphasis is on sharing of ideas and information to create a 'shared vision'. In learning organisations 'managers will become researchers and designers rather than controllers and overseers' (Crainer 1997:237). In *The fifth discipline fieldbook* (1994) Senge et al. provide five key concepts upon which organisational learning should be based.

Systems thinking

Systems thinking is a way to describe the 'forces and relationships that shape the behavior of systems' (Senge et al.:6). System archetypes can help managers to recognise repetitive patterns, particularly where problems recur. It is also incumbent on managers to understand the limits on a system's growth and development.

Personal mastery

Senge et al. advocate personal mastery as a way to allow individuals to expand their personal capacity in order to 'create the results we most desire.' In the learning organisation, the aim is to facilitate an environment which encourages its members to achieve their own goals in a way that is compatible with the organisation's espoused goals. Consequently, people 'live life from a creative rather than a reactive viewpoint' (Crainer 1997:238). The aim is to get people to continually learn to see current reality more clearly. Senge et al. argue that creative tension will be generated when there is a difference between vision and reality, a difference which stems from their learning and this, they suggest, will stimulate further learning.

Mental models

Mental models are the ways that people learn how to reflect on and clarify the world around them. As Senge et al. recommend, managers in a learning organisation should attempt to recognise the power of these patterns of thinking. By doing this, they will improve their ability to realise change.

Shared vision

Shared vision is created through 'building a sense of commitment by the group' (Senge et al. 1994:6). Indeed, as Sadler (1995:128) believes, 'where there is a genuine shared vision people become capable of outstanding achievements and significant learning takes place because of strong motivations.'

Team learning

Team learning has been identified as crucial in all other improvement initiatives. Senge et al. also advocate teamwork based upon dialogue and discussion. As they explain, it requires 'transforming conversational and collective thinking skills, so that groups of people can reliably develop intelligence and ability greater than the sum of individual members' talents' (Senge et al. 1994:6).

However, as Crainer (1997:239) points out, dialogue is exploratory by nature, whereas discussion is characterised by 'the opposite process of narrowing down the field to the best alternative for [a] decision to be made.' As he advises, they are complementary, but 'the benefit of combining them only comes from having previously separated them.' However, he warns that most teams do not see this distinction, so the desired benefits are less likely to be produced.

Senge et al. believe that a 'deep learning cycle constitutes the essence of a learning organization' (1994:18). Specifically they suggest that certain things must accompany it:

1. Three new skills and capabilities:

   ▪ Aspiration
   ▪ Reflection and conversation
   ▪ Conceptualisation

2. New awareness and sensibilities
3. New attitudes and beliefs

The learning organisation also needs a structure, a 'shell' or 'architecture' in which to occur. This has three facets:

1. *Guiding ideas* are the visions, values and purposes which are clearly articulated and understood by every person.
2. *Theory, method and tools* provide learners with something they can work with.
3. *Innovations in infrastructure* create a learning environment.

In addition, Mayo (1993) suggests the following characteristics for a learning organisation:

1. The use of the word *learning* is frequent.
2. The development of people is now the main objective for managers.
3. Those at the lower levels are consulted as a matter of course.
4. Feedback becomes a norm both upwards as well as down.
5. Learning models are decided by users.
6. Accidents or mistakes are events from which much useful learning can be generated; blame and covering up are out.
7. People see learning as being more important than status.
8. Information is shared.
9. The organisation is accessible and seeks to continually benchmark itself against best practice.
10. Spontaneity and informality are encouraged.

Crainer, who provides a commentary on *The fifth discipline fieldbook*, describes it as 'a tumbling spree of ideas and suggestions' (1996:203). However, he also believes it is notable for the simple message, 'Learning is positive; learning is good.' This is something that Senge and his coauthors stress with enthusiasm and energy. Crainer does point out the concept of learning is not without problems. Specifically it is a process which takes time and dedication. Certain obstacles must be overcome:

1. Managers may be unhappy with what is seen as dilution of power.
2. There is a need for flexibility and risk taking; something that many will be uneasy about.
3. The process is intrinsically uncertain. Uncertainty is traditionally anathema in most organisations which prefer to operate on the basis of predictability and certainty.
4. Every person will be expected to accept more responsibility.
5. New skills are required, which will take time, effort and money to introduce.
6. Trust is the keyword, something that may not come easily, especially to mangers not steeped in such an attitude.

12.2.1   Other writers on organisational learning

The most notable beneficiaries of Senge's work have been Chris Agyris and Donald Schon, who together wrote a book in 1978 called *Organisational learning*. In it they drew attention to what they termed single- and double-loop learning. Single-loop learning is where an organisation achieves what it set out to do. Thus there will have been instructions which went unquestioned. Double-loop learning is where the instructions (assumptions and beliefs) are questioned. The outcome may be very different than was originally anticipated. The significance is that single-loop learning is suitable for conditions of stability where routine is the norm. Double-loop learning is more appropriate to situations where rapid change is occurring.

In addition to these forms of learning, Agyris and Schon also proposed what they called 'deutro-learning'. This is the learning that accompanies investigation into systems the organisation uses to detect and correct errors.

Argyris's subsequent books have built upon the 1978 text. *On organizational learning* (1993) compares what he terms 'espoused theories' and 'theories-in-use'. The former are the beliefs, values and assumptions that are stated to exist, often by senior managers. The latter are the actual beliefs, values and assumptions that people 'draw upon when acting'. He advocates that an organisation should encourage managers to involve themselves in socialisation to the extent that they are aware of what theories are in current use. They can then attempt to align the organisation's espoused theories with the current theories. The alternative, he believes, will be to create a situation of conflict, error and resentment. These conditions could very well be extremely detrimental to the organisation's wellbeing.

Finally it is worth alluding to the work of Andrew Mayo and Elizabeth Lank. Their book, *The power of learning: a guide to gaining competitive advantage* (1994), explained that using this method is a long-term process. They also warn that in a rapidly changing market it is important to learn quickly. Failure to learn quickly will mean that an organisation gets left behind. Their main conclusion is that the employees of any organisation are as valuable as all other resources such as plant, equipment and buildings. If the people are trained and encouraged to learn, their ability to positively contribute is increased. This is not a tangible transaction that can be shown on the balance sheet. The result is that people are sometimes forgotten. They warn that forgetting people has dangerous consequences and advise organisations to make the following endeavours:

1. Attempt to study the magnitude of the hidden costs which result form failure to learn.
2. Measure what is added by allowing teams and individuals to learn and create.
3. Make the transfer of learning to newcomers as easy as possible.
4. Provide data which shows the comparison of costs and benefits of learning in relation to other techniques.

12.3 LEAN PRODUCTION AND MANAGEMENT

Many of the principles of Japanese management have already been described. One company is particularly noteworthy because of the system of management it has set up. This company is Toyota, the car manufacturer. Its system for production has become synonymous with what is called 'lean production'.

The significance of the Toyota production system to lean production is described by Womack, Jones and Roos in their seminal text, *The machine that changed the world*: 'After World War II, Eiji Toyoda and Taiichi Ohno at the Toyota Motor Company in Japan pioneered the concept of lean production. The rise of Japan to its current economic preeminence quickly followed, as other Japanese companies and industries copied this remarkable system' (1990:11).

According to Bounds et al. (1994 :8), the primary purpose of this system is to 'eliminate unnecessary elements in production to achieve cost reduction and satisfy consumer needs at the lowest possible cost.' There are three subgoals:

- Quality control
- Quality assurance
- Respect for humanity

The key concepts of the Toyota production system are as follows:

1. Just-in-time (*Kanban*) uses signposts and labels.
2. Autonomation (*Jidoka*) whereby automatic stopping devices are used to stop defects.
3. Flexible working (*Shojinka*) accommodates variation.
4. Creative thinking (*Soikufu*) is encouraged in the workers.

The importance of lean production, as pioneered by Toyota, is that it appears to be applicable to situations other than car manufacture. The central principle, according to Morton, is that 'quality is not seen as an issue: it is taken for granted that zero defects is the aim' (1994:167). Quite specifically, the objective of lean production is to get more for less, hence the word *lean*. The term was coined by John Krafick, a contributor to the study described in *The machine that changed the world*. The original term was *fragile*. Krafick believed this was a good summary of the delicate balance that car producers must achieve in using less effort but still maintaining their consistently high levels of quality.

Lean production aims to achieve the desired output by using

- Less non-productive human effort
- Less manufacturing space
- Less equipment

It also aims to generate fewer defects and to provide an ever growing variety of products.

Toyoda and Ohno were inspired by the American manufacturer Ford. However, their experience at the massive River Rouge plant in Detroit was memorable not for what to do, but for what not to do. Their belief was that the mass production (Tayloristic) techniques used by Ford, although apparently efficient and economical, in fact produced vast amounts of waste. Ohno believed that making small batches (of stampings) would alleviate the need for storage, hence reducing costs. More crucially, the sooner a part could be fitted, the sooner any problems could be recognised. This reduces the amount of scrap or rework.

Ohno's belief in the need to eliminate waste (in Japanese the word is *muda*) became the main reason behind efficient production. What Ohno also recognised was that mass production techniques which he saw in America were producing waste because of the way in which the workers were treated. His view was that 'the workforce was now as much a short-term fixed cost as the company's machinery, and, in the long-term, . . . an even more significant fixed cost' (Womack, Jones and Roos 1990:54). Thus, by only concentrating effort into making people conform to the needs of the machines and systems, and having single repetitive tasks which were usually deskilled, workers were not encouraged to contribute to the reduction of waste. In short, Ohno believed that the only way to achieve a better system than Ford's would be to have very skilled and motivated workers that could respond to 'rapidly changing demands' (Morton 1994:154).

Womack, Jones and Roos state that an organisation which is instituting lean principles has two organisational features (1990:99):

- *Transfer* of the maximum number of tasks and responsibilities to those workers actually adding value.
- *A System* for detecting defects that quickly traces every problem, once discovered, to its ultimate cause.

12.3.1   Four basic design methods

Having considered the work of Clark and Fujimoto (1988), Womack, Jones and Roos suggest there are four basic design methods which lean organisations employ in order to 'make it possible to do a better job faster with less effort' (1990:112).

Leadership

Every Toyota project has a *shusa* 'large-project leader', who possesses ultimate authority. The position requires true love in order to get the product into production. It also 'brings extraordinary satisfaction.' Although apparently similar to someone who coordinates mass production, the role of *shusa* is not the same. This is because the coordinator is usually 'in an extremely weak position to champion a project' (ibid.:113).

Teamwork

Teamwork in the lean production method requires a small team of functional specialists who are 'assigned to a development project for its life' (ibid.:114). The difference between a lean organisation and a mass producer is that departmental heads in the lean organisation recognise the efforts of their teams. According to Womack, Jones and Roos (1990:99) 'dynamic work teams [are] the heart of the lean factory' but they are not easy to build. Workers need to be taught a number of skills:

- To be multiskilled so that tasks can be rotated.
- To think and act proactively.
- To solve problems at source rather than waiting for someone above to do it.

These skills need to be complemented with a more enlightened attitude from managers. Womack, Jones and Roos (1990:99) have evidence to suggest that workers in plants trying to adopt lean production 'respond only when there exists some sense of reciprocal obligation, a sense that management actually values skilled workers, will make sacrifices to retain them, and is willing to delegate responsibility to the team.' The advice that follows is relevant to any quality improvement initiative: 'Merely changing the organisation chart to show "teams" and introducing quality circles to find ways to improve production processes are unlikely to make much difference.'

Communication

Most people believe communication is important, but many find it difficult. In lean production it is the *shusa's* job to ensure that all those involved in the project thrash out all the potential problems at the outset. Disagreement is accepted as being a real issue, but one which must be confronted. Individuals are expected to 'sign formal pledges to do exactly what everyone has agreed upon as a group' (ibid.:115).

Simultaneous development

Simultaneous development means that those who actually produce components are working hand in hand with the designers. Because of the intense communication and understanding of each other's needs and requirements, there is no need for them to wait until all the designs and specifications are complete. Womack, Jones and Roos point out that lean producers can get their products into production in half the time it takes mass producers.

Perhaps even more significant, realised Ohno and others at Toyota, is that in car manufacture the producer's objective is assembly. The vast majority of the components are actually made elsewhere.

Mass production ensures that all the parts come together at the right time; cost and quality are achieved by 'integrat[ing] the entire production system into one huge, bureaucratic command structure with orders coming down from the top' (Womack, Jones and Roos 1990:58). Legal contracts are created by commercial agreements and specifications, and business is often awarded on the basis of cost. Ohno wished the 'assembler and suppliers could work smoothly together to reduce costs and improve quality, whatever formal, legal relationships they might have' (ibid.:58). Specifically, suppliers should be able to contribute to improvements in the end product.

12.3.2   Five lean principles

Womack and Jones have written a more recent book, *Lean thinking* (1996). It contains five principles that explain how to apply lean production techniques within any organisation.

Specify value

Value can only be determined by customers who are provided with what they want, at a specific price, and at the right time. Womack and Jones believe 'the critical starting point for lean thinking is value' (ibid.:16). Consequently, it requires: 'a conscious attempt to precisely define value in terms of specific products with specific capabilities offered at specific prices through a dialogue with specific customers' (ibid.:19).

Identify the value stream

The value stream is the 'set of all the specific actions required to bring a specific product . . . through the three critical management steps of any business' (ibid.:19). There are three actions:

- Problem solving during the stages of concept to production.
- Information management during the stages from order to delivery of product.
- Physical transformation of raw products, using labour, to produce something the customer values highly.

According to Womack and Jones, any process can be analysed into three types of activity. First come the activities that 'unambiguously create value', obviously not a problem. But the other two kinds do present opportunities for improvement. The second type of activity is 'found to create no value but to be unavoidable with current technologies and production assets' (ibid.:20). They can only be improved by addressing the technology and assets. The third type of activity 'create[s] no value and [is] immediately avoidable' (ibid.:20).

Flow

The value stream should allow every part to contribute to value creation. Womack and Jones (1996:24) believe it is important to consider 'the real needs of employees at every point along the stream so it is actually in their interest to make value flow.' This, they suggest, requires a 'rethinking of conventional firms, functions and careers, and the development of a lean strategy.'

Pull

Womack and Jones believe this occurs when customer demands can be satisfied more quickly or more immediately. The ability to design, schedule and make exactly what the customer wants just when they want it means 'you can throw away sales forecasts and simply make what customers actually tell you they need' Womack and Jones (1996:24). Thus the customer is pulling your products or services rather than the provider having to push them onto indifferent customers.

Perfection

Perfection is something we should all aspire to but probably believe is impossible. According to Womack and Jones:

> As organizations begin to accurately specify value, identify the entire value stream, make the value-creating steps for specific products flow continuously, and let customers pull value from the enterprise, something very odd begins to happen. . . . Suddenly perfection, the fifth and final principle of lean thinking, doesn't seem like a crazy idea. (1996:25)

Micklethwait and Wooldridge (1996:271–74) agree that Japanese organisations have used these principles to devastating effect. In advocating the use of lean production, Womack, Jones and Roos state with absolute conviction that:

> Our conclusion is [that] Lean production is a superior way for humans to make things. It provides better products in wider variety at lower cost. Equally importantly, it provides more challenging and fulfilling work for employees at every level, from the factory to headquarters. It follows that the whole world should adopt lean production, and as quickly as possible. (1990:225)

Finally, it is interesting to note that the respected quality management writers and commentators, George Binney and Colin Williams make a pun on the word *lean* in the title of their book, *Leaning into the future* (1995). The principles they advocate are not based upon the Toyota model, but their ideas are by no means unsympathetic. Indeed it is their combination of leading and learning which they use to provide advice on how organisations must start 'leaning into the future.' Their main tenets are shown in Figure 12.2.

LEADING FROM THE TOP	LEANING INTO THE FUTURE	LEARNING FROM THE BOTTOM
Leader as hero Knows the answer; inspirational; wills others to follow	**Forthright and listening leadership** Combines assertive leadership with responsiveness to others	**Leader as facilitator** Self-aware; enables others to realise their potential
Vision Clear and inspiring visions, which explain why clean breaks with the past are needed, energise people to change; the answer is 'out there'	**Seeing clearly** The energy to change – and to develop existing strengths – comes from seeing clearly where the organisation is now and what possibilities are open to it	**Awareness** People change as they become more aware of their own needs and their interdependence with the world around them; the answer is 'within'
Dirve Change is driven through by determined individuals who plan carefully and minimise uncertainty	**Working with the grain** Individuals shape the future by combining clear intention with respect and understanding for people and organisations; they work *with*, not across, people's hopes and fears	**Release** Effective leaders release the natural potential of people and organisations to adapt to change and are prepared to live with uncertainty
They are the problem Individuals see the need for change in others	**All change** Leaders encourage others to change by recognising that they too need to shift	**We need to change** Change starts with me/us
Training People are taught new ways of working in extensive training programmes	**Learning while doing** Most learning takes place not in the classroom or training session but as people *do*, as they interact with others and reflect on their experience	**Reflection** People learn when they step back from day-to-day tasks and reflect deeply on their thoughts and feelings
ORGANISATIONS AS MACHINES	COMPLEMENTARY 'OPPOSITES'	ORGANISATIONS AS LIVING SYSTEMS

Fig. 12.2 Leaning into the future by leading and learning (Reprinted with permission, from *Leaning Into the Future* by George Binney and Colin Williams 1995:10; Published by Nicholas Brealey Publishing Ltd, Tel: (0171) 4300224, Fax: (0171) 4048311.)

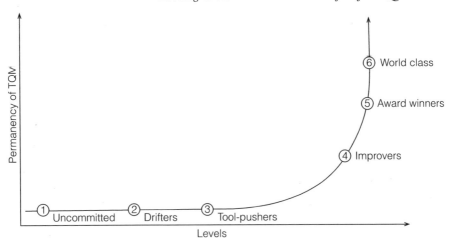

Fig. 12.3 The six levels of TQM adoption (Adapted from Lascelles and Dale 1993:285)

12.4 BECOMING WORLD CLASS: PRIZES AND AWARDS

Lascelles and Dale (1993) describe their research into TQM at the UMIST Quality Management Centre, along with ideas from some of their colleagues. They believe that TQM has six different levels of adoption (Figure 12.3). Each level comprises different characteristics and behaviours manifested by an organisation on its journey towards TQM. Here are the main features.

Level 1: uncommitted

Lascelles and Dale suggest this level includes those organisations implementing quality management solely as a result of customer pressure. Consequently, they will regard quality management as having no value; it will be carried out unwillingly. Indeed, as Dale, Lascelles and Boaden state, 'It is not given priority in terms of either managerial time or resource-allocation. Little or, more probably, no investment in quality management education and training will have taken place, and senior managers are reluctant to take responsibility for or get involved in improvement activities' (1994:118).

Uncommitted organisations are unlikely to develop anything more than maintaining registration. Merely achieving QA will produce less benefit when compared with a quality effort that is given higher status. Dale, Lascelles and Boaden stress that uncommitted organisations 'are dinosaurs belonging to another age' (ibid.:119).

Level 2: drifters

Drifters are organisations that 'have been engaged in a process of quality improvement for up to three years' (ibid.:119). An organisation at this stage will have expected much from their TQ initiative, and will be disappointed that all their problems have not been solved in one go. Dale, Lascelles and Boaden (ibid.:120) describe how an organisation will have some or all of the following characteristics:

1. Quality improvement is regarded as a programme rather than a process.
2. Quality management has a low profile and priority, and its implementation is limited to certain areas only.
3. Any improvement initiatives are little more than cosmetic, and efforts to produce teamwork are superficial.
4. Departments do not cooperate but persist with infighting and politics.
5. No real changes in corporate culture have been made since the start of the TQM initiative.
6. Senior managers are not showing a united front in their support of TQM.
7. Not everyone has been told what is going on.
8. A fear of failure pervades the organisation.

Such organisations are likely to continue with their efforts to improve, but in a half-hearted and haphazard manner. They will 'drift form one programme to another in a stop-start fashion' (ibid.:120). As a result, no real benefit will be likely. They need to concentrate their effort so it places 'quality improvement within a strategic business framework' (ibid.:121).

Level 3: tool-pushers

Tool-pushers are organisations which will have had experience of quality improvement for between three and five years. They will have introduced QA and have moved on to TQM in an enthusiastic fashion. They will be using a selection of quality management tools and techniques (Chapter 11). However, they will often use them in a superficial way and 'discard them once the novelty has worn off' (ibid.:121). The tools become blamed for being ineffective; little or no benefit is derived.

Dale, Lascelles and Boaden provide some characteristic features of tool-pushers (ibid.:122):

1. They are in search of the latest quick fix; quality prizes are regarded in a similar way.
2. Not all the management team are fully committed, which will be apparent to those at the operational level.
3. The management style is reactionary.
4. Meeting short-term targets is seem as the main objective, quality can come later.
5. Improvement is not widely practised.

Level 4: improvers

Improvers are organisations that are getting better. Figure 12.3 shows that improvers are actually on the upward part of the curve. This will have taken time and effort, and will have required everyone to commit themselves to long-term cultural change.

Dale, Lascelles and Boaden state that organisations at this level will be experiencing 'real impact on business performance'. Improvers have the following characteristics (ibid.:123)·

1. Coherent strategies and policies ensure that sustainable improvement is occurring.
2. Error prevention and problem solving are part of everyday activities in all aspects of the organisation.
3. Every employee is committed and is being encouraged to contribute by education and training.
4. Benchmarking is being carried out to ensure that lessons from proven performers are being transferred.
5. Quality improvement champions have emerged.
6. Hype, distrust and cynicism have been replaced by cooperation, teamwork and open communication; managers are trusted and supported.

It would seem that improvers have little to fear. However, Dale, Lascelles and Boaden suggest that, on the contrary, too much may depend on the champions; if trading conditions suddenly change, the gains will soon disappear. This was the fate of several organisations in the early nineties. Consequently, improvers must 'manage and coordinate quality improvement across entire streams of processes – the point at which quality improvement becomes total' (ibid.:124).

This, they suggest, will take between five and ten years. It will mean that an organisation is operating TQM in a way which will start to fulfil its expectations of potential benefits. It will also enable the organisation to move towards the next level, where any gains can truly be demonstrated.

Level 5: award winners

Award winners can consider themselves to have reached a point where 'quality improvement has become total in nature' (ibid.:124). They are able to compete against other leading organisations in order to win a quality prize. Their motivation will be connected with prestige, but they will also want to confirm their quality management is comparable to the 'best'.

The Deming prize was the first quality award; it was created by the Japanese in 1951 and named after Dr Deming; it continues to be held in extremely high regard. The award ceremony is broadcast live on television. The Deming prize contains ten elements which aim to assess an organisation's capability to create improvement:

- Policy
- Organisation and management
- Education
- Collection, dissemination and use of information on quality
- Analysis
- Standardisation
- Control
- Quality assurance

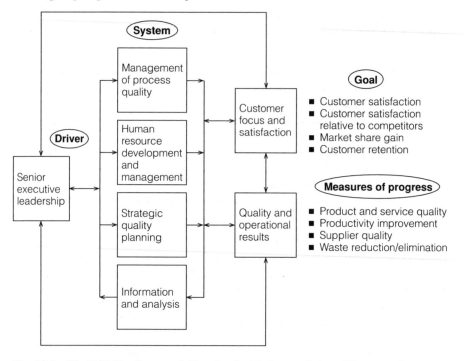

Fig. 12.4   The Baldridge framework (Reprinted with the permission of Butterworth–Heinemann, from Oakland 1993:150)

- Results
- Planning for the future

In 1989 the Deming prize was awarded for the first time to an organisation outside Japan, the Florida Light and Power Company. In 1987 the American government had set up an equivalent of the Deming prize, the Malcolm Baldridge National Quality Award. It aims to promote the following attributes:

- *Awareness* of quality as an essential element in competitiveness.
- *Improvement* in general understanding of the requirements for achieving quality excellence.
- *Fostering* of information sharing on the successful implementation of quality strategies and the potential benefits that can be gained.

According to Oakland (1993:150), Baldridge provides a 'universally accepted framework' for measuring an organisation's progress in implementing TQM. Figure 12.4 shows it has four basic elements: driver, system, measures of progress, and goal.

A European equivalent is the European Quality Award, launched in 1991 by the European Foundation for Quality Management in Paris. This foundation, set up in 1988, now includes some of Europe's 'leading' organisations. Intended to stimulate interest from all organisations, the award is made to those which have given

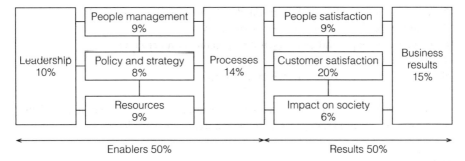

Fig. 12.5 The European Quality Award (Reprinted with the permission of Butterworth-Heinemann, from Oakland 1993:151)

'exceptional attention' to TQM. It was awarded for the first time in October 1992 to Rank Xerox, whose remarkable comeback was described in Chapter 11.

Figure 12.5 shows the main elements of the award. It is a model for self-appraisal, which means that an organisation uses it to compare its ability according to the categories provided. When it reaches a sufficiently high standard it can enter itself for the award. As Oakland states, the European Quality Award 'recognises that processes are the means by which a company or organisation harnesses and relates the talents of its people to produce results. Moreover, the processes and the people are the enablers which produce results' (1993:150). Points are awarded up to a maximum in each category. The different categories have different maxima, shown below in parentheses. The points in all categories are then totalled. A perfect total is 1000.

Leadership (120)   Leadership assesses a management's ability to provide the drive towards achieving excellence. It considers the following contributions:

- Visible involvement
- Ability to create an appropriate culture
- Recognition of people's efforts
- Support given
- Relationships with customers and suppliers
- Promotion of continuous improvement and quality values

Policy and strategy (80)   Policy and strategy assesses how effectively the management communicate the following items:

- Quality, value and concepts
- Relevant information in pursuit of the policies and strategy
- Business plans
- Actual communication
- Regular updating and improvement

People management (90)   People management emphasises how well the organisation is able to utilise the full potential of its most valuable resource, its people. It considers the following items:

- Alignment of people's skills with the development of business needs
- Training, recruitment and career progression
- Targets for people and teams (setting, agreement and review)
- Involvement and empowerment
- Communication between managers and workers

Resources (90) How does the organisation ensure all its resources are managed in support of policy and strategy? The following items are considered:

- Finance
- Information
- Materials from internal and external suppliers
- Whether technology is appropriate

Processes (140) Processes are the ways in which the organisation identifies, manages and reviews its ability to provide what the customer requires. This aspect is primarily the QA (ISO 9000) part of the model, so it considers

- Identification of the critical processes
- Management of all processes
- Measurement, target setting and review
- Innovation and creation in process improvement
- Process changes

People satisfaction (90) How well does the organisation satisfy its members? Both direct and indirect results are considered:

- Direct results from measures of employee perception
- Indirect results from predictions of how well the organisation can try to satisfy its employees

Customer satisfaction (200) Customer satisfaction is an essential part of the model; it judges

- Direct results in terms of measures to produce customer satisfaction
- Indirect results from predictions of how an organisation can satisfy its customers

Impact on society (60) Impact on society considers how effective the organisation is at considering the needs of the local and national communities, and how well it achieves their satisfaction. It looks at

- Direct results of specific measures
- Indirect results from attitude studies on how the organisation treats a community's needs

Results (150) The actual output of the organisation is assessed. Its products or services are compared directly and indirectly to plans and predictions:

- Direct results in the financial bottom line, i.e. profitability or other such measures
- Indirect results from other non-financial efficiency or effectiveness measures

In 1990 the British Quality Foundation believed it was appropriate to offer an award for organisations operating in the United Kingdom. Although originally called the British Quality Award, it was renamed the UK Quality Award in 1994. This brought it into line with the EFQM model, regarded as exemplary. The UK Quality Award is based on exactly the same criteria as the EFQM award.

As Dale, Lascelles and Boaden explain, even though an organisation may not actually win an award, the fact they have reached a point in their TQM maturity where they can even realistically consider entering for one will indicate that 'quality improvement has become Total in nature' (1994:124). They suggest these organisations will probably display the following characteristics:

1. A 'leadership culture' with total involvement and commitment to improvement.
2. Organisational change is accepted as a matter of course.
3. Measurable benefits will be occurring.
4. Strategic benchmarking is carried out at all levels.
5. Powers of decision making have been passed onto lower levels, and mangers are now acting as facilitators instead of commanders and controllers.
6. TQM is regarded as an essential method by which customers (both external and internal) are delighted by what they receive.

Level 6: world class

World class is the highest level of TQM adoption; it is achieved by those organisations which, possibly having achieved an award for quality, are searching for ways to 'identify more product and/or service factors or characteristics which will increase customer satisfaction' (ibid.:125). World class organisations will not be content to be regarded as 'the best', because they are aware that competitors will be attempting to catch up. The effort towards continuous improvement will be unceasing.

Until recently it was accepted that only Japanese organisations could rightfully be called world class (there are now some Western examples). The Japanese expression *miryokuteki hinshitsu* 'quality that fascinates' is germane to world class organisations. It requires a complete paradigm shift to have occurred in the organisation, its values, its customer commitment and its enthusiastic dedication to remain among the best. Failure to do this will mean loss of status, and more significant, a potential loss of competitive advantage, hard to regain quickly.

12.5 LOOKING TO THE FUTURE

It is almost trite to say that organisations must be forward-looking. Although any organisation may look back to past glories, it is vital to secure a future. However, in the modern world of organisations and business, nothing can be taken for granted. The development of products, innovation and increased quality are inevitable.

Japan continues to lead the way. One of the most successful books of recent times, *Competing for the future* (1994), was written by two Western academics, Gary Hamel and C. K. Prahalad. Their message was that organisations should continually aim to reinvent themselves by creating visions. Thus, rather than thinking about the short term, organisations should think innovatively and creatively. As a result, they should create new products around this vision. Many examples exist of organisations (and individuals) who created visions then went on to provide goods or service to satisfy newly created demand. It is accepted that others will follow the example provided, but although they can enjoy an overwhelming competitive advantage, they will possibly be able to sell at a higher price. The additional proceeds can be used to develop newer and more innovative products or services for the future. These innovations will be vital to any organisation which seeks to remain among the best.

SUMMARY

The techniques and methods described in this chapter are applied by any organisation that desires to go beyond TQM. The continual effort to improve will be firmly established, and employees at all levels can take advantage of one or more of the following:

1. The learning organisation encourages and facilitates learning by all its members. This will help the organisation to make a radical transformation followed by continual improvement.

2. Lean management is a set of techniques associated with the company Toyota. They are equally applicable to any organisation. Lean production (management) aims to eliminate waste at every stage of a process. There are various types of waste, all of which can be reduced and ultimately eliminated. This will make the process more efficient and cost-effective.

3. Attempting to become world class demonstrates that an organisation is reaching for the highest level of quality. However, the real value of attempting to become world class is that, by striving for the accolade, an effort to improve will have been stimulated.

Various awards provide criteria and methods of self-assessment to help orgaisations go beyond TQM: Baldridge (America); EFQM (Europe); UK Quality Award (United Kingdom).

QUESTIONS

Discussion

The ideas in this chapter may seem radical, but they have proved very successful in manufacturing. How can construction apply them to achieve similar success?

Individual

If you analyse any process there will usually be waste (*muda*), in terms of material, and especially time. Think of something simple that you as an individual have done recently. Using a flowchart, consider how you can achieve the task more effectively by reducing wasted effort.

(a) How much more effective will it be?

(b) Use your result for a single person to estimate the saving on a similar task performed by many people in an organisation.

Display your findings as though they are going to be presented to an audience; use bullet points, diagrams, short statements and recommendations.

FURTHER READING

Binney, G. and C. Williams (1995) *Leaning into the future: changing the way that people change organisations*. Nicholas Brealey, London.

Bounds, G. M., L. Yorks, M. Adams and G. Ranney (1994) *Beyond total quality management: towards the emerging paradigm*. McGraw-Hill, New York.

Flood, R. L. (1993) *Beyond TQM*. John Wiley, Chichester.

Hickman, C. R. and M. A. Silva (1986) *Creating excellence: managing corporate culture, strategy and change in the New Age*. Unwin Hyman, London.

Lascelles, D. M. and B. G. Dale (1993) *The road to quality*. IFS Publications, Bedford. Chapter 10 on becoming the best is particularly interesting.

Mayo, A. and E. Lank (1994) *The power of learning: a guide to gaining competitive advantage*. Institute of Personnel Development, London.

Morton, C. (1994) *Becoming world class*. Macmillan, Basingstoke.

Prescott, B. D. (1995) *Creating a world-class quality organisation: 10 essentials for business success*. Kogan Page, London.

Pumpin, C. (1993) *How world class companies became world class*. Gower, Aldershot.

Senge, P. (1990) *The fifth discipline: the art and practice of the learning organization*. Doubleday, New York.

Senge, P., A. Kleiner, C. Roberts, R. B. Ross and B. J. Smith (1994) *The fifth discipline handbook: strategies and tools for building a learning organization*. Nicholas Brealey, London.

Womack, J. P., D. T. Jones and D. Roos (1990) *The machine that changed the world*. Rawson Associates, New York.

Womack, J. P. and D. T. Jones (1996) *Lean thinking: banish waste and create wealth in your corporation*. Simon and Schuster, New York.

Zink, K. J. (1997) *Successful TQM: inside stories from European Quality Award Winners*. Gower, Aldershot.

CHAPTER 13

QUALITY MANAGEMENT IN THE CONSTRUCTION INDUSTRY: PRACTICAL APPLICATIONS

OBJECTIVES

- Provide empirical descriptions of how QA (ISO 9000) has been implemented in a number of large construction organisations.
- Present the experiences of quality managers.
- Indicate how some quality managers establish their authority.
- Describe the problems and obstacles that some quality managers have encountered.
- Explain how QA has helped to create cultural change, hence improvement.

13.1 CONTEXT: BILL'S STORY

One of the most lively and forward-thinking quality managers I met during my research was Bill. He is what many would call a character. His career spans many jobs and industries. He spent a long time in manufacturing before coming to construction. During our conversations, he described one of the most common problems that seems to beset the construction industry. Many times he would offer suggestions and back would come the reply, 'But Bill, you don't understand, this is construction, we're different, you can't apply quality management in the same way as manufacturing.'

Initially this rankled him. He was viewed as an outsider. As a result, his influence was limited because he hadn't been in the industry since boyhood. Bill's experience everywhere else had taught him to give the customers exactly what they wanted. This he had found was invariably achieved by clearly identifying their needs. It was then essential to ensure that the necessary tasks required were carried out efficiently. Bill wholeheartedly supports the adage, so often quoted about quality management, Get it right first time. As he explained:

> If you don't do this, it's more expensive for you. This represents a potential loss of profit, something that either you bear or try to pass onto the customer. But even worse, there is the problem of the image you present. If you don't do things right first time, customers are hardly likely to be left with a glowing impression.

Bill believed this applied to any process, no matter how simple or how complex. So when he was faced with opposition from those who saw quality management as something that doesn't apply to construction, he looked at what was being achieved without it. The results hardly indicated an industry which could regard itself as perfect:

> To be honest it was so easy to see that improvement was long overdue. Go onto any construction site and within a couple of minutes you'll see why. Materials being wasted, people wandering around seemingly with no idea of what they are doing, and a degree of confusion from management which is frightening.

Bill also stressed that construction is an industry that currently considers itself lucky to make a couple of percent profit, if at all. As he quickly decided, this was an industry that needed quality management as a matter of urgency, and the way to deal with those who resented him as an outsider was to ask the following sort of questions:

> Show me how you make sure you can get things done in the most efficient way? Why do you do it like that? What are the procedures you use to govern your processes? What measures do you use to ensure you are reducing waste, avoiding unnecessary cost, or making improvement?

These questions were usually met with evasion, blank looks, or those who told him, 'Its always been done like this.' Given his personality – he said himself he would take no bull – Bill knew that people who answered his questions could be hanging themselves out to dry. However, his belief is based on the principle that if you are going to institute improvement, it will need to be by teamwork. This requires support, even from those who initially resent you:

> Getting people to see your point of view is the main objective. It's certainly not easy in any industry, but especially in construction. The aim must be to get them to commit themselves, the 'hearts and minds' thing. You can leave the detractors till later. They take up too much time and energy. Besides, in my experience of managing quality, once the rest of the group are on board, they will usually isolate and destroy them.

Bill strongly believed that construction is no different from other industries. He has made good use of the techniques he learned elsewhere, particularly at a world class machine-tool company. As will be described later, he is a 'change agent'. From the research that I have carried out, Bill is very much an exception in construction. Most quality managers, who I believe are the critical ingredient of change management using quality, are not as forward thinking. However, this does not mean their results will not be comparable in the end; they will just take a bit longer.

The rest of this chapter and the next will describe their efforts at implementing quality management in construction.

13.2   INTRODUCING QA TO CONSTRUCTION

For many construction organisations the introduction of 'formal QA' using a quality system such as ISO 9000 was not voluntary. Most organisations implemented QA because they read newspaper predictions that the government, through its agencies, would in future demand that contractors tendering for work should be accredited to the British Standard for quality management BS 5750 (Chevin 1991:48; Walsh 1995:92).

BS 5750 was published in 1979. Following the launch of the National Quality Campaign of April 1983, a decision was made to provide an example to industry. This was to demonstrate the 'government's determination to enhance the status of standards and quality assurance in the United Kingdom' (Morrison 1994: 43).

Many large contractors who carried out work for government bodies, such as the Property Service Agency (PSA) or the Department of Transport (DOT), sought to become registered with BS 5750 in the late 1980s. Their motivation was fear; if they failed to register, they believed it would preclude them from even being considered for work which had in the past significantly contributed to their turnover. There were government grants provided to help pay for consultants. This initially alleviated some of the contractors' concerns about what would be required. But using consultants to develop an organisation's quality system is now regarded as not the way to do it.

My own interest was stimulated by reading the claims for QA made by its advocates compared to the problems raised by its detractors. The first article that I consciously remember was in 1986 by John Pateman. He asserted how QA was necessary because it meant that costly mistakes could be avoided:

> At the very heart of quality management is the principle that 'prevention is better than cure' or, put another way, 'It always costs less in the long run if you get it right first time'. This of course is also a fundamental precept of QA. Quality management, however, takes it to the limit by allying it to the maxim 'Money is the most important material we deal with'. (1986:18)

Others concurred and a number of books were published to help construction organisations with QA (Ashford 1989; Foster 1989; Building Employers Confederation 1990; Chartered Institute of Building 1990; Construction Industry Training Board 1990; Duncan, Thorpe and Sumner 1990; Griffith 1990). All essentially had the same message: QA using a quality system such as BS 5750 (as it was then known) is likely to bring benefit to any organisation which implements it. As a result, the whole construction industry would benefit. Some writers described the likely benefits. Typical were Hughes and Williams (1991:7), who gave the following eleven advantages of a 'systems approach':

1. Improved communications and efficiency.
2. Checking of work and avoidance of unnecessary and costly errors, failures and expensive remedial works.
3. Documented proof that work has been executed in compliance with the document and specifications.
4. Easier implementation of client changes.
5. Precise clarification and quantification of the effects of such changes.
6. Easier identification and quantification of delays and claims.

7. Completion on time.
8. Reduced maintenance period remedial works.
9. Provision of as-built records
10. Possible reduction of insurance premiums.
11. Improved competitiveness and marketability of services.

These claims, which were made without direct reference to any empirical data, indicated that QA using a quality system such as BS 5750 was 'too good to miss'. However, the late 1980s and early 1990s saw a number of articles appear which seriously questioned the use of QA as a means to ensure organisational improvement (Chevin 1991; Hamilton 1991; Parker 1993; Billingham and Stewart 1993; Ridout 1994). One of the chief concerns of these commentators was the amount of paperwork that would be generated by a formal system of procedures. There were also references to BS 5750 having been written for manufacturing, tacitly implying it would be inappropriate for the construction industry.

The claims and counter-claims that appeared did not tell the reader what was actually going on. My interest was to do precisely that. I wanted to find out what was really happening now that quality management had been introduced to construction.

13.3   MY RESEARCH OBJECTIVES

The debate that accompanied the introduction of quality management into construction had provided a phenomenon ripe for research. But how? It seemed my role as a researcher would be to prove or disprove the claims from either the advocates or the detractors. I therefore decided to use the traditional method of formal structured interviews or postal questionnaires. However, I felt somewhat uneasy about using this sort of approach, usually known as positivism.

My concerns were fuelled by the doubts that using this methodology would only give me superficial answers. I wanted more than simply statistics, which although seemingly impressive, might be of little use in finding out what QA really meant. What I wanted to know was, How were people in construction organisations managing (or not) as a result of the apparent imposition of QA?

My experience, and the rationale for choosing a qualitative methodology (participant observation), are described in greater detail elsewhere (McCabe 1997). There are many books that cover the issues for researchers who study organisations; among them I would recommend Bryman (1988), Easterby-Smith, Thorpe and Lowe (1991) and Shaffir and Stebbins (1991). Watson (1994) provides an account of his research which is both illuminating and provides excellent insights into the day-to-day management of an organisation. Not only did it inspire me, it provided an exemplar for the data I will present.

My decision to study the work of quality managers as 'agents of change' was influenced by the realisation that these people were the keystones. Without them, it appeared, initiatives to bring about organisational change using quality management would be extremely unlikely to succeed. As a result, my methodology has allowed me to present data which describes how practitioners deal with quality management on a day-to day basis in their own words. Thus what follows is no longer theory, it is how quality management is really being applied in construction.

13.4   THE FIRMS IN MY SAMPLE

I do not name any of the firms that contributed to my research, some twelve in all, because I gave them a commitment of confidentiality. The names of all quality managers have been changed so they cannot be identified.

The firms, all of which operate in the private sector, agreed to cooperate following a letter or telephone call. All carry out a wide variety of work in the construction sector, although one carries out only large-scale civil engineering. They operate either at regional (i.e. Midlands) or national level. As a consequence, over a three-year period, I visited offices and sites as far apart as Leeds and the South Coast, and from South Wales to the Wash.

My choice of firms was mostly a matter of judgment, but also a little bit of luck. The judgement was to contact large contractors, firms I understood were most likely to be implementing QA. The luck came in getting to talk to someone sympathetic. In a couple of cases my contact followed a recommendation. Bill was someone who I actually met by chance when visiting a site with another quality manager.

The final number, twelve, is not one I specifically chose. I contacted fifteen firms and twelve were willing to collaborate. I would suggest that one of the reasons I was able to translate cold contacts into research collaborators was my explanation that I wanted to see things as they really happened. On more than one occasion the quality managers I regularly visited complained at the number of poorly drafted and extremely simplistic questionnaires they received with monotonous regularity. I would advise any researcher to get out there. My experience was that people in industry are only too happy to cooperate if students or researchers are willing to spend time and effort at the sharp end.

13.5   OLD FACES FOR A NEW JOB

As Chapter 6 described, there are several things that must be done to achieve QA. It is debatable as to what causes the biggest change. However, one of the specific requirements of the quality standard ISO 9000 is the need to appoint a management representative. This role is crucial. The person appointed (invariably called the quality manager) will be given the task of achieving registration of the organisation in accordance with the British Standard.

Quality managers became the focus of my study. Despite the need for senior management support and commitment in order to set the lead and give the right example, quality managers – normally middle managers – will determine the success of any quality management initiative. This is because, as I have repeated many times, those who operate at the lower levels of any organisation control what goes on. In effect, they produce the products or services. What a quality manager must do is to assist and encourage them to work to their fullest potential.

Consequently, quality managers are those who will put a quality initiative into practice. They should also help to ensure that improvement actually occurs. As Chapter 6 described, quality management using ISO 9000 can be achieved by force or threat. However, in the long term, this sort of strategy is not going to be sustainable.

One of the main concerns of those I talked to was their lack of understanding about the role of a quality manager before they took on the job. Typical was Brian, a self-professed old hand:

> I sometimes think that if I knew then what I know now, there's no way that I would have accepted the job as quality manager. When it was put to me, it sounded so simple. I was told that this thing QA was required and that it would probably take a couple of months.
>
> It had been a bit slow and I was between contracts at the time, so it seemed like a good idea. I was also told that QA was no big thing. It was just to ensure that all the existing procedures were written up and a few extra ones created. Unfortunately it was not as simple as that. The system as it stands today has had to be virtually created from scratch. That has taken a lot of effort, something old hands like me find very strenuous. As I describe myself, I'm an old face in a new job.

This interesting view sums up what it means to be a quality manager. On several occasions people expressed the view that QA was perceived either as a part-time role or as a task for someone with nothing better to do. In two cases the job had deliberately been given to someone who was a couple of years from retirement. This was often interpreted by the rest of the employees as sending out the message that QA was not very important. But Ray was approaching retirement and he saw it like this:

> The age thing can be looked at two ways. Some might see the QA job as a soft touch. Something where you can ease off. That, in my opinion, is rubbish. It is difficult, stressful and requires a lot of patience, especially when you are explaining why you will not write their procedures. I've been around long enough to know that people will do things if you take the time and trouble to describe what they must do, and then help them to do it. This takes experience and respect. I personally think getting the QMS [quality management system] up and running requires an older head.

Ray told me about a sister company where a newly recruited graduate had been given the job of achieving registration, but this led to problems:

> There was no holding him back. The trouble was S had read all the books but he had never worked on site. He thought that by telling people what he wanted written it would get the system implemented. He had forgotten the cardinal rule of this job. That is, if you want to implement a quality system, the only people

that can do it are those who actually carry out the operations. They must feel ownership of the procedures. Consequently, when he had his first external audit, no one was adhering to them. He tried being high and mighty, but the attitude of many was to tell him that if he wanted their support and respect, he'd have to earn it. As I said to S, this game is about getting people to do what they think will help them, not you.

This seemed like common sense. It was not surprising that during the course of my meetings with quality managers the word *ownership* was repeatedly mentioned. Ownership seems to be agreed upon by all who advocate QA. Without it, quality managers come to be regarded as policemen of an imposed and unpopular system. The very reason for implementing QA is what carries the problems. Colin explained:

> There is no doubt in my mind that this industry went for QA because of the fear that without it big clients would exclude them from tendering. Thus the top management of all the big contractors said to themselves, 'Right we'll get QA regardless of what anyone thinks.' There was to be no consultation. That happened here. It meant that I could tell people to do whatever I wanted or else they would be out. However, I knew that would make me as popular as a rat sandwich. I preferred to keep the people out there on my side. They will be the ones who have to use the system.

And Bob confirmed:

> The need to have the support of those at operational level is crucial. However, having our managing director tell everyone that we were going for QA, and anyone who didn't like it knew where the door was, did not help me. Initially I felt that my biggest task was to convince them that I was not a lackey [stooge] who was doing their dirty work.

These are the credibility problems faced by quality managers in the construction firms I visited. In particular, they need authority which will give them support from the majority, and ability to extract compliance from the minority who resist. This is not something peculiar to construction. The use of authority was described by the German sociologist Max Weber (1864–1920) when he distinguished between coercion and authority. Coercion causes action because of fear, whereas authority is based on willing compliance.

Weber identified three sources of authority: rational, traditional and charismatic. Rational authority occurs where there is a system of rules which require adherence because they are believed to be the most rational (efficient) way of doing things. Senior managers often argue that such a rationale is the basis of a quality system. Thus, the system's procedures cannot be disobeyed, and the quality manager can use this as a reason for demanding compliance. Traditional authority is where there is custom and practice. A person commands respect and can exercise authority because their job perpetuates the established way of doing things. In QA terms, this is what many of the quality managers told me was their approach; the system merely documented existing practice.

Charismatic authority is where a person is able to get others to follow or act in a certain way because of their personality and ability to inspire. This was described under leadership in Chapter 9. In relation to QA, a quality manger will use this form of authority to convince others that the system is something they can use to benefit themselves. It is up to people to buy into the belief that QA is positive, and they must ensure the procedures are used as a way to improve.

13.6   THE QUALITY MANAGER AS ENFORCER

From the research I carried out, many quality managers were put under pressure to get their organisation certified to BS 5750 as a matter of urgency. In many cases the easiest thing was to use the formal authority that senior managers had vested in them. They used the 'do it or else' form of persuasion. Coupled with managers' doubts about their suitability for QA, this has caused problems with implementation. Ray sums it up:

> I'll admit that this job is not what I thought it would be. I don't know if I would have taken it. At the time there seemed little choice. My real interest is in being out there [onsite] doing things, not going around telling people that they should follow this or that procedure.
>
> What makes the job harder is that no one thinks the system is well written. This wasn't down to me. A consultant was employed to do that. Consequently the procedures bear no relation to the way things are really done. I would like to improve the system, but that would mean dismantling it and starting again. It's something that I've raised with my boss. His answer is that we got registration so why complain. So I have to use the threat of the MD's wrath if someone doesn't comply. My job is simply to police the system. Its not a happy state of affairs.

13.7   WRITING PROCEDURES INTERACTIVELY

The predicament Ray describes is that quality managers are required to police a system which has no support from the users. They must then use the threat of punishment to ensure compliance, which despite being an essential prerequisite to initial and continued registration, is not the way to engender enthusiasm. A much better method of managing quality is to use traditional authority. The system therefore does not try to do anything other than document existing practice, whether it be good or otherwise. But this is not always straightforward, as Bob describes:

> I was lucky enough to be in another company and see how QA can be done really badly. By that I mean where procedures and manuals are invented by outsiders [consultants]. The QA manager had no chance. So when I came here to get registration, I was determined not to do the same. The trouble was that, as a

newcomer, it was going to take some time to get to know what goes on, and who to really trust. I needed to get out and about. But after about ten days the MD started asking when the procedures would be written, as if I could do this myself. I told him what I intended to do. That the users would write them, and that the system would be based on existing practice. You should have seen his face! He's no fool, and realises that there are things done which shouldn't be.

He was telling me that there was no way we could document the way operations are carried out as they currently stood. He believed that the external assessor would not allow it. As I explained, this is not the way that they operate. By telling the auditor what we really do, and then by demonstrating we were actually doing it, was being honest. As I also argued, the exposure of shady practices would act as a basis for improvement on the way that those operations are carried out. He accepted this, albeit reluctantly. As it turned out, he was easier to convince than those who I really needed on board.

The site and office staff are, I believe, essential in that they need to write their own procedures. Some were easier to convince than others. There were one or two who told me in no uncertain terms that I was the QA manager, I should write the system. When I told them, fair enough, I'll record that you don't want to be involved, they thought I was joking. But when I sent them a copy of minutes which said this, they soon changed their tune. I'll admit that I wasn't really intending to go to higher authority, but they were not to know.

Once I could get the attention of those who carry out tasks which were to be covered by the system, I got them together. This was interesting. I told them to create a sensible and minimal set of procedures to cover them. Everyone thinks that their way is best, and that they can do it in about two minutes. These sessions sometimes got a bit heated. This is where I had to intervene. But after much argument, procedures started to emerge. But what was really good was that some of the dirty linen was aired. As I made clear, I had no problem with this. If that's the way it's done, then that's the way the procedure should be written.

There was usually a debate as to whether we should use these procedures which represented poor practice. As I stated at this point, if they wanted to institute ones that were representative of a better way of doing things, then fine. But I would not tolerate a situation where I was administering a false system. They [the actual users] needed to be sure that the new procedure was the one that would be followed. I let the majority rule on this one. You normally find that dissenters are put in their place. But this is not something I would want to do.

When I asked Bob how the system actually operated, he told me there had been teething problems. They were of the sort which caused no undue concern, and as he admitted, they 'represented some continuing debate.' He concluded, with a sense of satisfaction:

On the whole, the system is easy to audit. There is none of this nonsense of records being filled in the night before an external assessment. As I believe, the system is a developing thing. It can be continually improved. That is my real responsibility, to facilitate this continual change.

Like other successful quality managers I spoke to, Bob has undoubtedly found that by looking to the users to write the system his job was made much simpler. Indeed, as some stressed, the task can be enjoyable.

13.8   THE QUALITY MANAGER AS CHARISMATIC

Charismatic authority is most likely to serve the quality manager best in the long run. Bob would probably call himself just someone who looks to the group to tell him what is standard practice. However, as he did admit, the system can be continually improved. There are some quality managers who believe that real improvement comes not from writing better procedures, but from trying to make the transition to a more proactive and informal way of working. This process is the basis of TQM.

Making the transition requires someone to convince people of the benefits that come from an improvement strategy. Chapter 14 describes how this is being achieved in some of the firms I visited. But Martin, one of the charismatics, stressed that the system can still be retained:

> There is a danger that by advocating a less formal approach, you appear to be binning the system. I make it clear that QA is a foundation. What I really want from people is the ability to show initiative and imagination. Quality is about getting the right product to the customer. If they don't like what we provide, no amount of procedures will convince them otherwise. As I tell our directors, the QA system is to enable, not disable. Thus, if someone tells me that a procedure is actually stopping them doing a good job, I ask tell them to show me how the task can be achieved either more efficiently, or will give the customer better value. I will then get them to think about a new procedure, or alternatively as I increasingly advocate, show it in flow diagram form.
>
> The aim, I believe, is to get people to think for themselves. They shouldn't have to wait to be told. They also shouldn't be restrained by out-of-date procedures. This is a radical view. But I base my belief on the fact that all of our people are very dedicated. They are used to thinking on their feet. My job is to allow them to do that in a way which is flexible but not undisciplined. The biggest barrier, though, is senior managers. Most I've come across find it difficult to let go. They think that if you do that, there will be anarchy!

13.9   GETTING THE SUBBIES INVOLVED

Most texts which describe quality management usually make some mention of the need to involve suppliers in efforts to improve quality. In the vast majority of organisations there will be some services or components which are supplied by outsiders. Consequently, any improvement effort should involve them.

Many of the organisations which are now recognised as being world class rely on the dedication of suppliers (often known as partners) in order to achieve the high standard of the end product or service. In the construction industry there exists a situation where practically every event or task will be the culmination of many different parties. In carrying out my research of quality management in large contractors, this was very much the case. The efforts of these organisations were very much about bringing their own staff up to speed. However, there was recognition that subcontractors and suppliers should be encouraged to play an active part, not an easy task. Mike gave this description:

> The biggest problem we as an industry have is the number of firms that we have to deal with. The suppliers of materials is bad enough. But when you consider those firms that supply and fix, or just provide labour, it's a nightmare.
>
> Under QA we are supposed to be able to demonstrate that we have methods to adequately control these firms, both in terms of their selection and the product and/or the service they provide. Reasonable enough, you might have thought. I knew that when I consulted the approved list that we select from, there would be a fair few. But blow me, even I didn't realise quite how many we, as just one contractor, use. There are literally thousands. It's just a company Yellow Pages. Quite how you are supposed to control them, I don't know. In the long run though, if we as an industry are going to improve, we have to get the subbies involved.

This statement demonstrates the problem that most construction firms face when attempting to get their suppliers and subcontractors involved. The majority of their concern tends to be directed at the subcontractors. Suppliers, on the whole, tend to be fixed-location operations. Many are factory-type producers, so they have already taken on QA. Subcontractors pose a different problem.

Firstly, the number of small firms tends to be large (Hillebrandt 1984; Ball 1988; Harvey and Ashworth 1993), and the contracting system has encouraged their use. They allow large firms to subcontract when extra work is available. Thus, the number of small specialist firms has increased to service this demand when it occurs. These specialist firms have to remain small in order to keep their overheads minimal during lean times. This means that all of them attempt to be included in the lists of approved firms used by principal contractors.

The second problem comes from the small size of the subcontractors. Smallness often leads to minimal administration. The owners are generally involved in day-to-day operations. When a principal contractor asks for QA, it can be an additional and very unwelcome burden. However, for the quality managers I talked to, the fact that so many subcontractors are small proved to be a mixed blessing. As Ray believes:

> Everyone accepts that the small subcontractors are the lifeblood of this industry. They give us the ability to expand and contract, which is vital in the current market. The problem is that, when we look to them to develop quality management, there is a problem.
>
> For instance, under QA we need to get quality plans from all the subcontractors. But I know that some of these guys are working at full stretch. Their work is not a problem. In fact, some of the best work is produced by the 'one man and his dog outfit'. The trouble is that their paperwork leaves a lot to be desired. It is a difficult situation. Do we demand that they do the extra

administration just to satisfy QA? The danger is that either these firms will go elsewhere to avoid the hassle. Or worse, their work will suffer because of the additional burden. My attitude is that all that counts is the end result. The paperwork is secondary.

A site manager told me how he had dealt with the QA concerns of a small subcontractor. In the process he actually assisted the firm to get more work:

> It was a bloke I'd known for years. He did bloody marvellous work. But like a lot of the firms we use, his admin side was a bit ropy. I, like a lot of the site managers, didn't mind helping him with this. He got us out of problems when we needed him. The least I could do was to help him. Because, believe me, he was in a terrible state.
>
> So when he told me that our buying department had written to say that, unless he got QA, he wouldn't be used in the future, I was livid. I knew there would be no point in arguing his case. The reply would be that he is no different to everyone else. So I sat down with him after work one night and went through the standard procedure to carry out his operation. In fact, it didn't take long. I got the secretary to type it up on his headed notepaper. You should have seen his face the next time he came to the site when I showed him this. He was over the moon. As I explained, in future, all he would need to do was to follow this procedure. He could also sign and date each stage as he completed it. It was simple but effective. As he has told me since, he now submits this with tenders to other firms, and that it has helped him to get contracts.

This account was an excellent example of the way in which QA can be used by small subcontractors. The important point is the site manager recognised how he needed to support the subcontractor in his effort to maintain his standard of work. Indeed, I was able to witness many other examples of dedication to maintaining good relationships. As another quality manager put it, 'Its simple stuff really. We scratch each other backs. If we don't, everyone loses out.' However, the number of small firms involved still presented a problem, and for every one of these success stories there were plenty of other firms who either ignored QA, or treated its introduction with contempt. It was a situation that often required a strong hand. Much depended on a firm's ability to produce good work before the introduction of QA. With questionable firms, QA represented an opportunity to get rid of them. However, this did not always happen. Colin was somewhat despairing:

> Something that really annoys me is that we get rid of those who have always caused us problems. They may be cheap but we all know that in the long run they are more expensive. But lo and behold, these firms get some sort of buckshee registration for QA, and they're back. This is what gives QA a bad name. The site staff tell me, 'We don't care if they've got QA, they're still rubbish.' When I question the buying department, I'm told that cost is the main priority. This is one of my crusades. We should be willing to pay the bit extra to get the better subbies. This is what I believe has been the approach in manufacturing. It's a lesson this industry should follow.

It might seem as though the contractors I visited prefer only to work with large suppliers and subcontractors. This is not always the case. Indeed, the experience of those I interviewed indicated that although some of their firms are reasonably large, they are not necessarily better. Mike's account is typical:

> Certainly there is more chance that the bigger firms will be more likely to have 5750 [BS 5750]. But it's one of those ironies that, despite the fact that my job is to sell QA, I don't always believe that just because a supplier has QA it will mean they produce a superior product.
>
> Take M's. They are one of the biggest firms around here. They are very competitive and have a good stock. But their service and attitude stinks. They have got QA, but if we could get the materials cheaper elsewhere we would. We are more or less forced to go to them. This is not quality. I would prefer to use small merchants because you are more likely to get personal service. That's what I believe quality is really about.

This attitude also applied to the experience of subcontractors. Don was a regular companion on site visits; he told me:

> There is no doubt that some of the worst problems we've had, have been from subcontractors with QA. For instance, that job we visited in Brighton. If you recall, the reason it was running late was due to C, the cladding contractor. You would think that by using a firm as big as them, you would be all right. Their paperwork is fine. But the product and their ability to install it is not.
>
> This is a common problem. Firms like that get QA because they think it will impress. They even have a full-time QA manager. Between you and me, I think he is useless. He can only just about keep up with the paperwork. In terms of getting the factory to produce, his influence is zero. The firm then subs out all the fixing. You never know who you'll get turning up. I'm not joking, some of the blokes that have been sent were just not up to the job. They didn't have an effing clue.

So why do they stay with this firm?

> Because they are the only one with the capacity to deal with the sort of jobs we do. If there was a choice we'd use others.

The most obvious difficulty is that, despite the exhortation to use only firms which have quality assurance, there can be no practical guarantee of their standard of work. That the industry relies on subcontracting means the actual work often falls to individuals. As one QA manager described:

> The big firms are cutting costs. So when they get a contract they sub out. That's all right, but the blokes who actually do the work end up getting a pittance. This is where QA falls apart. It's something we all contribute to. Until we get back to the situation where there is some control over labour, we don't really have any hope of improvement.

This was a recurrent theme. Many interviewees described how it was in the old days. QA was regarded as a poor substitute for the traditional methods of doing things. There seemed to be a feeling that until the environment changed, there was little hope of trust. The ability of individuals to help each other had been undermined.

Subcontractors are required as standard to submit quality plans for each job. Quality plans represent an opportunity to sort out problems before work begins. But as a site manager explained:

> Like all these things, a lot depends on the attitude of both parties. I want my job to run smoothly. To do that, I need the subcontractors to cooperate. If they need some assistance, then its in my interest to give it. Providing what they require is not going to cost too much.
>
> But if by the time they get to me, they've been squeezed by our surveyors, then they couldn't give a toss. QA doesn't help. In fact, their attitude is 'more bloody paperwork'. Some are worse than others, but I do sympathise. They are just trying to make a living.

This last comment summed up the desire of many I talked to. Everyone wants to do a good job, and QA can sometimes help. More important, though, is the ability of people to cooperate. This may be facilitated by a less formal approach such as TQM. It is what many of the firms I visited are attempting to achieve (Chapter 14).

13.10 SENIOR MANAGERS AND QA

It is crucial to have senior managers who are committed to the introduction of quality management. Without such commitment, the initiative will either fail to start or it will quickly founder. I have never heard of an organisation that implemented quality management without support at its senior levels.

Nevertheless, I did come across an alarming number of examples where commitment was ambivalent at best. This made life difficult for the QA manager. It also sent out the wrong signals to the rest of the workforce. Bob captures a quintessential expression of unease with senior management:

> They gave me all the support that I expected in the lead-up to registration, including the usual huff and puff about how it would produce tremendous change. But as soon as the certificate was on the wall, they lost interest. I've tried to keep them involved, but to be honest, I've given up. Their attitude to any problem is, 'You're the QA manager, you sort it out.'

Another difficulty appears to be the belief by senior managers that QA does not directly concern them. George, another QA manager, puts it like this:

> They are quick enough to condemn others who don't adhere to the system. But they see QA as being about procedures. That ISO 9000 only involves following procedures for operations. The senior managers here think they are above all of that.

These attitudes make it very difficult for QA managers to get compliance with the procedures. More worryingly, it makes further improvement using TQM almost impossible. Don describes how a lack of commitment from senior managers may very well cause people to give up their efforts to develop the system:

Our directors made a lot of positive noises about what QA was going to do. They said that education and training was essential. But when things got tight, training and day release were the first thing to go. I complained, but got the 'it's unfortunate but necessary' argument. I persisted in telling the MD that if we were to stand any chance of real improvement leading to TQM, how crucial it was that QA is underpinned by investing in training.

I was taken to one side by my immediate boss, who said, 'You're being seen by some as becoming a bit of an irritant. The board's attitude is that all that is required from you is to get them to follow simple procedures. People know what their job consists of. What more training do they need?' Needless to say, after that I took the stance that if that's all they want, then that's all they'll get.

For some of the quality managers I came across, the biggest challenge was in getting their superiors to lead by example. These quality managers looked to the long term. They refused to accept the arguments that were advanced as to why short-term aims came first. This requires confidence, and Pete, who I would describe as a charismatic manager, told me:

If you agree to even the smallest compromise, they'll have you by the balls. You will find that every time you need even the smallest expenditure, they [senior management] will say, 'It's only QA. It's not that important, it can wait.' Once this happens, you're stuffed. The word soon gets out. Before you know it, the attitude in the office and particularly on sites is the same. You can't use the argument that senior management require adherence to procedures. They'll say, 'We hear that the top boys don't think it's that important.'

Pete did not lack confidence. He believed that if you are to be a quality manager who genuinely wants to create improvement, then you must become a 'change agent'. He stressed that you need 'the courage of your convictions.' This, he believed, was certainly important in terms of QA. But when it came to TQM, he said:

Its absolutely fundamental to be able to drive the people at the top as much as those at the bottom. This is not easy. They [senior managers] expect that all the change is going to come from below. They find it a real culture shock when I tell them that they have to get off their backsides and lead from the front.

The next chapter gives more detail on the potential problems of obtaining support from senior management. However, before trying to achieve that, the quality managers who were attempting a transition to TQM stressed the importance of consolidation as a way to maintain momentum.

13.11   MAINTAINING MOMENTUM

One of the problems that seems to affect organisations which have achieved accreditation is the belief that the hard work is over and everyone can relax. This is amply described by Mike:

There seems to be an initial euphoria. It's like you think, 'Great, we've got it [ISO 9000].' The problem is everyone thinks that the hard work is over. The people upstairs [senior management] are the worst. They had a reception and invited the press to celebrate us getting QA. They thought that the quality system would run itself. Those at the site who should use the procedures to carry out operations, also think that once the certificate is on the wall their input is over. As I have had to keep explaining, this is just the beginning. The worst is yet to come.

Mike is alluding to the fact that, post registration, there must be a realisation that QA is here for good. It must be continuously implemented. He went on:

Getting the certification is one thing. Keeping it is quite another. The experience here is similar to others, in that fatigue sets in quickly. You get the 'can't we forget QA, and get on with the real job' argument from the line managers.

And Bob commented:

Maintaining momentum is the key to success. At the very least, you need the system to be adhered to as it was intended. If you, as a QA manager, let anyone off the hook, so to speak, and allow them not to have to comply, you'll be finished. Everyone will come along wanting to do the same. I've heard all the excuses here, and in my previous firms. We didn't think that we'd actually be expected to use the procedures. The lads don't like keeping records. And so on. Eventually they'll try anything. My granny isn't well. My budgie has died.

No, if you want to keep the system going, you have to be absolute on compliance; no exceptions. If there is something wrong with a procedure, I have no problem with it being rewritten. In the long term it will make for an easier life. Once people get used to using the basic procedures, it becomes second nature. That doesn't happen overnight. It takes a lot of hard work to achieve. In the long term it will also enable you to start to think about the next step, improvement techniques and TQM.

One of the biggest problems is the visit of the external assessors. In every organisation I visited, there was annoyance at the quality of assessors. It was especially the case that in the early days of QA, they were found to be wanting. This was apparently due to the large take-up of BS 5750. The accreditation bodies found that they needed extra staff, some of whom left a lot to be desired. This did not help those who were trying to ensure the assessments were consistent. As Don explained:

To say that the assessors were variable is an understatement. For the first year you'd get a different one every time. This should be fine in theory. The trouble is that each one would have their own pet desires, and want to see more use of this or that. It was very disruptive. Once we got someone on a regular basis, things got better. A lot of our people got a bit fed up along the way.

Further criticism was levelled at the audits themselves. They were seen merely as a nit-picking of the paperwork, with little concern for whether improvement could be demonstrated. I learned that, in the trade, assessors are known as seagulls. QA manager Adrian explained why:

> Third-party assessors? We call them seagulls. The reason for this is that they come to your office for the day. They fly around, squawking a lot. They eat all the food you provide for them. Then they shit on you!

I asked him to elaborate:

> Some of the non-compliances they pick up are pathetic. The date missing, or a number out of sequence. The sort of administration errors that will always occur. Our people are doers, not filing clerks. It doesn't really help me to convince people that the system is worthwhile.

All my interviewees agreed that the external bodies which award ISO 9000 should ensure they are better able to assess the organisational systems currently in operation. They stressed how assessors should begin by looking for compliance, but with genuine consistency. However, in the long run, they should be looking for measurable improvement. Dave put it like this:

> Maybe we, as QA managers, are expecting too much. I don't think we are. But QA using 5750 or 9000 has been a real mixed bag. I suppose without it, we wouldn't be here. Personally, I think it is just the starting-point. If we can get QA right, we can start to move on towards something more ambitious. Then we can think about real improvement. Until then, we must get our houses in order. The only way to do that is to use QA consistently.

Dave is not exceptional. He is typical of those I encountered, managers who believe that TQM in the construction industry is both possible and desirable. As the next chapter explains, this will require much greater effort than was needed for QA.

SUMMARY

This chapter has described the experiences of twelve quality managers working in large construction organisations. Their ability to create success as a result of implementing ISO 9000 has been hard won. It has required much effort and the active contribution of all concerned. Extracts from taped interviews have shown there are many hurdles, but positive change can nonetheless be created. The following issues are worthy of note:

1. Quality assurance is often seen as a bureaucratic imposition, and its implementation will be unwelcome.

2. External assessment is a particularly difficult part of QA.

3. The quality manager is normally the person responsible for organising the development and writing of the quality system. This must be achieved by facilitating contributions from the people who will use them. The quality

manager should not draft them in isolation then impose them upon the rest of the workforce.

4. The system must reflect what goes on, whether or not it is good practice. This will represent the starting-point of improvement.

5. It is highly beneficial to involve people from outside the organisation implementing QA, e.g. suppliers and subcontractors.

6. The support and commitment of senior management is essential.

7. There is no time to relax once registration has been achieved; the hard work must continue.

8. A good QA system will be the foundation for further improvement initiatives such as TQM.

QUESTIONS

Discussion

1. A common belief of organisations is that QA will not help them to do their work any better, it merely documents what goes on. There is also much criticism that QA creates a paper necklace. Although this may be true, how much do you think QA can assist organisations to ensure their customers are satisfied?

2. Why is it that when you shop in high street stores, you can usually rely on their reputation, QA is not part of your requirements? Can construction learn from this?

Individual

Draft a one-page job description for a quality manager; address the following questions:

(a) Given the importance of the position, what skills should an organisation look for when advertising for a quality manager?

(b) If an organisation wanted to conduct interviews, what sort of questions should be put to the candidates?

FURTHER READING

Cheetham, D. W. and J. Lewis (1993) Implementing quality plans – the role of the subcontractor, in *Proceedings of the Ninth Annual ARCOM Conference*, Oxford University, pp. 124–39.
Cheetham, D. W. (1996) Are quality systems possible? in D. A. Langford and A. Retik (eds) *The organization and management of construction: shaping theory and practice*, Vol. 1. E. & F.N. Spon, London, pp. 364–78.

Construction Industry Research and Information Association (1996) *Quality management in construction: survey of experiences with BS 5750*. CIRIA, London.

Gray, C. and R. Flanagan (1989) *The changing role of specialist and trade contractors*. Chartered Institute of Building, Ascot.

Hodgkinson, R. (1995) Factors affecting a firm's ability to make a successful transition to quality management, in *Proceedings of Practice Management for Land, Property and Construction Professionals*, Liverpool John Moores University, pp. 133–40.

Hugill, D. (1995) The relative merits of introducing a formalised quality management system, in *Proceedings of Practice Management for Land, Property and Construction Professionals*, Liverpool John Moores University, pp. 125–32.

McCabe, S. J. Rooke and D. Seymour (1995) Quality managers and cultural change, in *Proceedings of the Eleventh Annual ARCOM (Association of Researchers in Construction Management) Conference*, University of York, pp. 452–59.

McCabe, S. (1996) Creating excellence in construction companies: UK contractors' experiences of quality initiatives. *TQM Magazine*, 8 (6), 14–19.

Naoum, S. G. and M. F. Mustapha (1994) Contractor's perception towards the application of quality assurance in the UK construction industry, in *Proceedings of the Tenth Annual ARCOM Conference*, Loughborough University, pp. 465–74.

Seymour, D. and S. McCabe (1994) Is it possible not to take sides? Researching the impact of BS 5750 in construction, in *Proceedings of the Tenth Annual ARCOM Conference*, Loughborough University, pp. 454–64.

CHAPTER 14

THE JOURNEY TO TQM: CONTRACTORS' EXPERIENCES

OBJECTIVES

- Describe how some construction organisations have begun the transition to TQM.
- Explain the preliminaries for TQM transition.
- Show that TQM relies on people's active involvement and commitment.
- Indicate that quality managers are vital for cultural change with top-down passion and bottom-up enthusiasm.
- Explain that world class status is ultimately possible despite its apparent remoteness.
- Emphasise that construction firms can learn from other sectors of business, regardless of size or function.

14.1   CONTEXT

The title of this chapter was inspired by the title of a book, *The road to quality* (1993), written by Lascelles and Dale. It was mentioned in Chapter 12, along with the various stages of TQM adoption. Figure 12.3 showed that the graph becomes very steep when improvement starts to occur. This indicates how moving towards TQM is not that easy.

The difficulty of making the transition from QA to TQM is described by Bill:

> Getting ISO 9000 is like riding a bike up a small hill. Once you get to the top, people think that you can freewheel. But you've got to keep everyone working consistently, like you're on a plateau. This is because moving to TQM will be like coming to a mountain. You know it will be hard and it will hurt.

Not all the organisations I visited had made the decision to move towards TQM. And those that had were following different approaches. This is not surprising. As many commentators suggest, the whole point of TQM is that an organisation implements whatever produces the best results in terms of meeting its customers' expectations. QA is differentiated from TQM in that there is a standard for quality management, ISO 9000, which contains clauses to be followed. Although standards exist for TQM, BS 7859: Parts 1 and 2 (1992), they are intended only as guidance on management principles and improvement methods, respectively.

This chapter presents the experiences of those who have the courage to embark on what is going to be a long and arduous journey towards TQM. The success of their efforts in the long term is hard to judge at present. The hope of these contractors is that, by emulating world class manufacturers and perhaps other world class organisations, they will be able to enjoy similar benefits, especially competitive advantage.

14.2   A CONCRETE STORY

This concerns one of the contractors who had experienced traditional problems with concrete: the finished dimensions would not meet the specification. The ready–mix firm was occasionally given the wrong formula. Sometimes the strength didn't match up to expectations. The result was expense and loss of time. George, the company quality manager, recalled 'the incessant sound of jackhammers on our sites, something that everyone took for granted.' He explained this was something he had always believed could be cured, given enough effort. However, in the past, no one had considered it to be either necessary, or indeed possible. His appointment to the role of quality manager was the opportunity to do something:

> The first thing I did was to get those who were involved in the whole process together. I also deliberately included the ready-mix firms we normally used. As I told them all, I had done some investigation on the problem of concrete. My initial estimates were that, as a company, the loss due to rework was in the order of a quarter of a million, and that was a good year.
>
> I explained that the purpose of the meeting was to consider all the problem areas and their likely causes. As I stressed, the objective was to use Pareto analysis to identify which were the causes of the largest problems in terms of cost. This seemed to cause great amusement. I then told them about the principle of Pareto, where 20% of the problems cause 80% of the recoverable cost.
>
> This got them interested. But the next stage was the hardest part. Getting some constructive debate was not easy. For years all of the departments had blamed each other. In turn they all blamed the ready-mix firms, who in turn blamed us. It's no wonder we couldn't get the concrete right, all the effort was going into slagging each other off. This is the biggest problem that affects this industry – the blame culture.

This was the first of several meetings George held. And getting the various parties to talk without arguing was not straightforward:

> You have to keep them from trying to settle old scores. There is no point in that. What I want is to get everyone thinking as a whole, not as individual departments. I also want everyone to think creatively about the future.

Eventually the meetings began to produce tangible results. The most important was an agreement on how all the parties could communicate more effectively in order to avoid future problems:

> You'll hear that word used all the time, communication. Everyone will say, 'If only we could improve communications, we could sort out potential problems.' In the case of our concrete problems, I got each of the parties to agree methods by which they could more clearly communicate their requirements to each other. As I told them, it doesn't matter how formal or informal the methods are. Procedures are necessary only if the users think they will help do the job. In most cases, all that is required is to understand what the other side needs, and the time required for action.
>
> The reality is, I believe, that just getting all the parties together is the main objective. Apart from the inevitable finger-pointing, my experience has been that, once this happens, people will start to sort out their problems. They are not bloody children after all. We are all professionals. That is the beauty of TQM, it's common sense. But try explaining that in advance.

How effective was it in resolving the problems with the concrete? George was clearly delighted:

> Its early days. But so far so good. The sound of jackhammers is not quite so widespread. Indications are that the costs have started to reduce. The main thing is that I can use this as an example to attack the problems with other trades.

14.3   SETTING UP A QUALITY CIRCLE ONSITE

Bill's story is similar to George's. Bill set up quality circles based on what he learned in manufacturing. It involved bringing together all the people on a number of sites. The first stage is crucial, he believes. The period of initiation will determine whether the circle succeeds or fails:

> You have to be strong at first. You have to lead them by being a bit of a bully. They [site personnel] see you as interfering with the most important thing – getting the job done. So to start off with, it is difficult to even get them to turn up at all, and if they do, on time. Further, they are often not prepared to think or be disciplined in the way they conduct themselves. Thus you have to use a bit of muscle to influence them. The best way is to tell them that non-attendance will be viewed in a poor light by senior management. This is usually enough to at least get them there, in body, if not soul.

As Bill explained, the belief of many who work onsite is that meetings are only held because you have to 'kick someone', usually one of the subcontractors. The idea of turning up to think about how to problem solve is regarded as an alien concept. Bill also thinks there is a belief that 'if you're not kicking someone, you bollock them to work faster.' This, he further explains, provides that the next stage in the development of the quality circle: the need to develop an alternative to the established way of thinking, a 'quality culture':

> Once I've got them there, I stress the need for the meeting to be conducted according to rules. I get everyone to agree that those who cannot be bothered to turn up on time are excluded. This means the door must be locked. If this happens to those who are late, they will be sufficiently embarrassed not to do it again. Further, it sends a message out that we as a group are serious, and we are not going to tolerate those who are not prepared to contribute.

Even this can be far from straightforward. On one site there was no lock on the door to the room where meetings were held. Bill asked who was going to get one fitted; there was no response. It took a while before someone volunteered to do it, and, when Bill asked how long the job would take, he got an evasive answer. Bill's response:

> That says it all. You can't even get a lock fitted in a definite period of time. You are supposed to be setting an example to those out there [site operatives]. Imagine what the client would think if they heard this.

Bill admitted it was a deliberate attempt to embarrass the group. However, it had the desired effect. When the meeting was held the following week, the lock had been fitted.

According to Bill, the first couple of months are the most difficult, largely because people start out by expecting to be told what to do. There is also the question of who will nominally 'lead' the quality circle. Bill believes in sorting this out before any further progress is made:

> I stress that this is a group, and every member is equal. They are all expected to contribute. There is always the question of who will be the chair and/or spokesman. On most occasions I've been involved, the most senior member of staff who attends assumes that they will take the role. The rest of the group will go along with this on the assumption that this person will tell them what to do. I deliberately force the group to accept someone to be leader who is not the most senior. It's a symbolic gesture. I want to show that seniority counts for nothing in quality circles. It will annoy the person who expected to be the leader, but I want people to feel that they can be open and honest. They will not do this if they feel the group is just an extension of the management hierarchy.

Having achieved this, the next stage is to organise the actual meetings. According to Bill, this is also something that people usually find difficult. The reason is that most people will not be familiar with working to an agenda. Even those who are, often find it difficult to keep to it. As he stresses, there needs to be some training. Also, he advises, someone has to be appointed to take the minutes. Action points need to be agreed, and group members need to realise they will be held accountable to achieve them. As Bill explained:

> This can take some time to get accepted. My influence will be very high at the outset. I accept that leadership from me is essential. However, after some time

the message will start to get through. The group begin to realise that they can achieve things on their own. By using statistical techniques and tools that I will have taught them, they can start to identify and solve problems.

I insist that they produce reports to senior management on site and at head office. Apart from obviously informing senior managers what is being done, and what is required of them in the future, they demonstrate that the group has a voice at the highest level. This gives the quality circle the feeling of esteem, and a sense of the importance of their task. The confidence that this gives will allow them to move onto more ambitious aims.

It is at this point, Bill believes, that he can withdraw. The group will be able to fend for itself; it no longer needs a facilitator (Bill frequently uses this term). When the facilitator withdraws, it may be seen as a sign of success. Besides, he stresses , 'there are plenty of others who will still need help.' But he does advise some caution:

There can be a belief of some who think that they can now solve every problem the company has. They must not abuse their ability to communicate directly to the board. I have to intervene to ensure that they stick to problems that directly concern them at this stage.

Part of Bill's philosophy is countering the belief that it is the major problems which must be solved first. This, he thinks, is fallacious. Large problems require large effort, and there is the danger that a solution may not always succeed. Failure at this stage will have negative consequences; it will undermine people's desire to try again. This is something Bill is extremely keen to avoid:

I get the groups to concentrate on small things. These are more likely to be within their influence, and provide some quick victories. This will be good for morale and gives enthusiasm to continue their efforts.

What examples of success had he witnessed?

On one site recently, they flowcharted the process of dealing with enquiries for technical information within the company [which is design and build]. As a result, they had been able to rationalise the process and save approximately nineteen minutes per query.

They didn't think that this was a particularly marvellous achievement, until I pointed out that until we do things radically differently, we have hundreds if not thousands of enquiries on each job. Thus the saving in time for each contract will be nineteen times that number. The total could be enough to make up for the time spent in the quality circle meetings. So any further solutions they generate will be free.

This is indeed an impressive indication of what quality circles can do. I asked Bill what he would do when he managed to set them up on every site and office. Was there any danger he would become superfluous?

Someone said that to me recently. I agreed that it is possible in the very long term. But my belief is that this indicates that I'm doing my job as a change agent. I don't see that situation occurring in the near future. This industry, and the company in particular, have a long way to go.

14.4 GETTING CLOSE TO THE CLIENT

TQM helps an organisation to get closer to its clients. The use of formal QA has helped, but it requires more than that. Gordon is making the transition towards TQM; he says:

> In the past we've all played the contractual game. They gave us a hard time, and we used the contract to hit them. It was almost enjoyable. But when you start to look at the costs, both direct in terms of lawyers, and indirect in terms of reputation, you see where it leads. Everyone loses out, apart from the lawyers.
>
> QA helped to tidy up a lot of things. It took a while, but we sorted out many of our problem areas. It allowed us to be able to stand above a lot of the nonsense. If an architect tells us something that is not true, it's odds on we've got a procedure that will cover our position. I know it still seems that we are playing the contractual game, but in this industry respect comes from being able to defend your position.
>
> That word, respect, is important. What I want is for us to build upon what we've achieved so far with QA. Call it what you like, but until the client starts to give us respect there will be no improvement. To be fair, in many cases we didn't deserve it as an industry. We were happy to behave as the ignorant builders. You know, all macho but no style. Now we, like many of the big boys, realise that if we can get closer to the client by being seen as part of the professional team, there are benefits for both sides.

When I asked him to elaborate on the sort of steps he had taken, he explained that it was 'simple stuff'. By this he meant that ensuring there were regular modes of communication. There was also the need to recognise that problems can be sorted out if they are 'caught early enough'. Overall, he stressed:

> I personally hate the expression, but it's about 'culture change' on both sides. In any relationship there will be a need to find the common ground. Previously it took the lawyers to establish where it was. Now we try to get there without them. One of the things I've suggested is to forget the contract. You should have seen some of the directors' faces, they thought I was mad. I explained that the amount of time and effort we put into 'managing' contracts, and produce so little at the end, there must be a better way.

Gordon admitted his suggestion came from something he had read about the way in which Japanese organisations do business. It was, he reflected, a 'step too far' at the time. However, as a means of stimulating debate, it certainly worked.

John's experience was similar to Gordon's; he was working on a project for a large financial institution:

Luckily it was a design and build job, so we were able to have influence from day one. However, it was obvious from the way they treated us initially that they thought we were not on the same level. As one of our marketing men said to me, 'They think we are scum.' I was involved in setting up the quality plan for the job. I told our senior managers this was not good enough, we should not tolerate their attitude. If they wanted us to do their work, they should understand our problems.

I recommended that the client's people come to the design meetings and regularly visit the site. Our guys were not very keen. The client was also very unsure, they didn't expect that we'd want them so closely involved. Apart from a few teething problems, the arrangement worked really well. I know that they [the client] now appreciate that construction is not what they seemed to think it was. They also realise we are not cowboys, which seems to be the stereotypical image of this industry.

The result is that we were able to give them something which they are delighted with. In fact, it's up for an award. The benefit for us is that we got the job done on time, and within a budget that they are happy with. We even made some decent profit. The best part is that they intend to use us in future, Apparently, they are now crowing about their partnership with us, like it was their idea.

This account confirms the belief that genuine improvement is both possible and very desirable for all sides. It indicates that, whilst the industry can improve its ability and image, educating the client is equally essential. It also shows that the concept of partnership works downwards as well as upwards.

14.5 USING SUBCONTRACTORS' EXPERTISE

Subcontractors are a major part of the construction process, so they must be involved in any improvement effort. However, as George explains, this is not as easy as some may think:

If you say to our staff, 'We must get our subcontractors to perform better,' you'll get no opposition. But if you say, 'How can we help them to perform better?' the answer will tend to be, 'That's not our concern.' There is a belief that subcontractors are there to be used and abused. Fair enough, some don't deserve any better. But I know from experience, that there are some bloody good firms out there. They've been hit as badly as us by the recession. So what do we do? Screw them even tighter. Some big contractors who hold their money for months, like they have to prove they deserve to be paid.

This is wrong. They do all our work. Therefore the end result, what the client sees is down to them. My own conviction is that if we can get our subcontractors to perform more effectively, they produce a better end result. In the process they will make more money, we get less problems, and the client will be happier. Sounds simple doesn't it? Try telling some of our people!

The huge number of subcontractors in each trade has proved one of the biggest impediments to greater involvement. This undoubtedly makes it difficult, as one quality manager told me, to 'separate the wheat from the chaff.' It also impedes those who wish to develop closer relationships with particular subcontractors. But as other sectors have shown, particularly manufacturing, it is vital to get suppliers to contribute to the overall process. This appears to require a huge change in the thinking of many construction workers. Don ventured some ideas:

> It's easy to always blame those below you. Subcontractors have frequently been the whipping boys. This is part of the culture. I believe that part of the process of improvement is to look hard at yourself. This is something that we as large contractors find difficult. We are too fond of saying, 'If only they would do this or that,' or 'We need to hit them harder with the contract.' We should be more willing to stand back and say, 'What can we do to make their life easier?' When you say this to some of our site managers, they look at you as if you've grown horns.
>
> I'm about to attempt something which has taken a lot of effort to achieve, but I hope will contribute to finding out what we can do to assist subcontractors. This is a questionnaire to those who work for us. It asks what they think we could do to help them produce their work more effectively and efficiently. There were many who were opposed to this. There were arguments such as 'What do we care what they think?' or 'Their job is to do what we tell them.'

Don firmly believes in developing closer ties with the people below as well as those above. There is nothing terribly sophisticated about this; a consultant told me the following story:

> I heard about this job in London where the client was very insistent on the plaster finish that they wanted. They asked the architect if it was possible. He (the architect) thought that it was going to be difficult to know how they could specify it. So the client went to the contractor who would probably get the job. The contractor said it was certainly achievable, but were also not sure how it would be possible to specify it so that they could be sure that their subcontractor would achieve exactly what the client wanted.
>
> This seemed like a stalemate until someone had the bright idea that maybe the best people to ask would be the those who would do it, the operatives. So the contractor got the best firm they use for plastering, to send a couple of their tradesmen to see the client. The tradesmen were then taken to the place where the client had seen the sort of work that they wanted. Apparently the tradesmen said, 'No problem, as long as certain things are done in advance.' And that was that. All the endless writing of specifications was avoided, most of which just cause dispute anyway.

This clearly demonstrates that the expertise in the 'system' lies at the operational level, something that Japanese organisations learned from Deming around fifty years ago. Martin, one of the more enlightened quality mangers, told me:

It's a sad reflection on this industry that we have tended to ignore what Deming argued. That is that quality is about the people. For too long this industry has treated its only resource, people, in the most despicable way. This is the reason that everything is subbed out. Contracts are awarded on the basis of price alone. Deming warned against precisely that kind of attitude. Well, we can see where the Japanese are, and where construction is.

This attitude reflects the despair that is often felt among those who are attempting to provide a new way of thinking. The construction industry is felt to place too much emphasis on contracting out responsibility. Among those I encountered, the advocates of TQM consider it essential to get beyond the slavish adherence to contracts. After all, trust and teamwork were both encouraged by the Latham Report (1994).

14.6 SENIOR MANAGERS AND TQM

Support from senior managers is regarded as crucial by advocates of quality management. Without their desire to get an initiative going, there is unlikely to be much commitment from employees at any level. This was described for the implementation of QA using ISO 9000. It is perhaps even more true for applications of quality management that go beyond QA.

One of the major concerns of the quality managers I met was the difficulty they often experienced with 'those above'. Dave, a quality manager for a large contractor, defined the potential to improve as 'a direct correlation of the input that those on the ground see from the leaders of the organisation.'

The question of leadership was described in great detail in Chapter 9. And it came up very frequently during my interviews. Quality managers are only able to achieve a limited amount unless the senior managers 'reach down'. When Bill set up his quality circle (Section 14.3) it was essential for the group to feel their efforts were being recognised at the highest level. But in many of the organisations I visited, the senior managers found it difficult to be considered as part of the team. George offered an explanation:

It is probably part of the way many of them made it to the top. By being aggressive and keeping information to themselves. It tends to be the 'I'm up here, and don't need to look back' attitude. The use of TQM needs to challenge this as much as getting those at the sharp end to take more responsibility.

Many of the quality managers found this a daunting prospect. They had usually been appointed to implement QA. Neither they, nor their superiors had envisaged TQM. Their ability to 'shake things upstairs,' as one quality manager put it, is limited unless they are willing to 'put their neck on the line.' However, this did not pose a problem for Bill:

I was appointed to improve. They [senior managers] thought that I would only be involved at the operations end. They got a rude awakening when I made it clear that I expected them to contribute. I'm sure they thought, 'Who does he think he is? He can't talk to us like that.' My experience has taught me that every person from the top to the bottom must be involved. I don't fear someone because they have a more senior position.

This attitude had allowed Bill to get to the 'decision makers'. Consequently, it meant that quality had assumed a higher status than many other organisations I visited. Bill explained that, despite his energetic nature, he needed others to 'champion' the cause. The most influential people to do this, he stressed, are those who are regarded as the leaders. One of the aspects that he regarded as being essential in this regard was to be able to communicate:

Too many of the senior managers in this industry think that you can communicate by memo. As I tell them, 'The only way to ensure support by those at the lower levels is to make them feel important. They are not going to feel special if the only time they hear from you is on a piece of paper.' They tell me, 'We're too busy to get out and about.' I reckon that most of what they do is deal with trivial administration that could easily be delegated. I'm not afraid to tell them as much. It is my firm conviction that leaders should be like figureheads. If they are stuck in the head office, no one will think of them in that way. It's a big challenge to get them to accept this principle.

Given that it's crucial to have commitment from senior managers, how would Bill advise people who found themselves in a position like his?

The biggest problem is awareness and clarity of objectives. If you as a quality manager are not sure about what is required, possibly because those who appointed you also weren't sure, then it will be impossible to establish what must be done, and when. My experience is that if the targets are not firmly established at the outset, if any difficult patches occur, the TQM will be jettisoned. You have to have the balls to tell them what their responsibilities are, right from the start. The trouble with quality management is that many of those who are appointed are not likely to challenge those above. Thus they get QA implemented. Even that is in a fashion. TQM is talked about, but that is all it remains – talk.

This was typical of Bill, always willing to have a go. But he was right. As the adage says, Train your pup and you have your dog. Train your boss and you have TQM. The only problem is that the bosses appoint the quality managers in the first place!

Finally, it is worth quoting Thomas (1995:137–38) on the relationship between senior managers and TQM. His advice is entirely resonant with what I was told onsite:

1. Senior management must fully understand the central philosophy and key concepts of TQM.
2. Senior management must be fully aware of the tremendous challenges in becoming a TQM organisation.

3. Senior management must be totally committed to this aim.
4. Senior management must act in accordance with the principles of TQM (i.e. walk the talk).
5. Senior management must understand that, in the vast majority of instances, major changes will need to be made in terms of organisational structure, culture and working practices.
6. In any TQM programme there is likely to be an initial honeymoon period before the critical challenges begin to emerge.
7. Certain people within the organisation, often powerful, will offer strong resistance to the changes that are necessary for TQM to succeed.
8. At some point it will be very much easier to abandon the TQM programme than to keep it going.
9. Everyone will have doubts about the viability of the programme at some point, even the 'quality freaks'.
10. Senior management must give continuous support to the programme and, through their actions, show that they are determined the programme will succeed.

14.7 CULTURAL CHANGE IN CONSTRUCTION

Those who attempt improvement initiatives are essentially trying for a change in culture. And to achieve that, people need to have 'bought into it,' says Dave, a quality manager. Senior managers do play a vital part, but those at the lower levels are essential for realising their wishes. Bob describes himself as a quality fanatic and his overall outlook is similar to Bill's:

You only have to see what other industries have done to realise what quality can do. Too many people in construction tend to pooh-pooh the experiences that come from outside. They must learn that some of the organisations we accept as being top performers now, were on the verge of collapse or bankruptcy. What they did was take on the principles of quality, and especially what the Japanese have done. By that, I mean they recognised that their people must be part of the process. This is the biggest challenge to construction. For too long people have been treated as an expendable commodity. I'm forever telling our directors this.

Don admitted:

When I first became a quality manager, I never expected the role to assume the importance I now believe it should have. That's not because I think it has to have rank or any of that sort of thing. No, it is because you are the person who will be responsible for getting the people to support the effort. This is not because you wave a stick, or threaten them. But because they see value in going along with it. It's the 'hearts and minds' stuff. That has been a huge challenge to me, but an even bigger one to our people at all levels. Too much effort in the past has been wasted on setting up meaningless systems and departments. Quality has taught me that, whatever we do in the future, we must put the people first.

Don continues to be ribbed by the site workers:

> As they say to me, 'Bloody hell, you're beginning to sound like a commie, all this looking to the workers thing.' I can see what they mean. In an industry which has traditionally been for the tough men, this philosophy sounds very radical. But look at the state we're in. Low profits, poor conditions and pay. What will be the future unless we get our act together? I'll tell you, my son is coming up to his A levels. If he asks for my advice, I'll tell him to get into something like manufacturing. That is something I would never have dreamt I'd say ten years ago.

Don's concern is heartfelt. He has spent all his working life in construction, and he feels the industry has lost many of the things which made it enjoyable. Don was one of the many who told me that, unless construction organisations decide to invest in their people, especially the young, the industry will not survive. He also felt the industry was 'too cut-throat', leaving little room for sentiment:

> You used to be able to say it was enjoyable working here. Not any more. There is too much emphasis on the profit margin. Everyone is too busy to enjoy it. That is the difference that I think would help to restore people's confidence and willingness to go the extra bit. At the moment, when you talk to people, the message is that they do their bit, but if they could get out they would. That's a sad reflection.

But some are beginning to redress the balance. George is one of them:

> It is not only vital, but extremely fulfilling to get people to be involved in the quality management efforts. I've set up a number of teams on our sites. The change in their thinking and attitude is remarkable. They are the ones pushing for more change, not me. My job is now to facilitate this. It's like a revolution, and the directors cannot stop it even if they wanted to. The bottom line is literally that we are making more money. We can achieve a better job for our clients with less errors and rework. These are things that give people a pride in the job, something that was fast disappearing.

14.8   SOME CONCLUDING THOUGHTS

It has been my privilege to be associated with a number of quality managers who formed an informal group at the end of 1994. The aim at the time was simply to talk to one another. As all admitted, the job of quality is a lonely one. It is a new role, often viewed as something which will interfere with normal working.

The coming together of the quality managers in this group was a way of providing mutual support for each other's efforts. Initially the discussion was limited to QA. Even this has caused some debate about what is required. But increasingly the group has focused on how to develop an ability to learn the lessons of TQM. What has been fascinating is to observe the different attitudes and approaches that the various quality managers have in terms of how improvement can be achieved in their respective organisations.

To summarise the situation in the group, I would say it is rather like Figure 12.3. Most participants are on the flat part of the curve. A couple have started improvement initiatives; they are beginning to move up the curve. For the rest, some difficult times lie ahead. Everyone is committed. The fact that quality managers turn up to share information is in itself a notable shift in culture. Many of the initial meetings discussed the problem of confidentiality. The discussion appears to have helped, for the issue is now successfully resolved. The meetings have also allowed the participants to deal with their own uncertainty, and they have gained strength through unity.

What the group of individuals have really achieved in their period of existence is to realise that their concerns are not unique. This gives comfort and the desire to share ideas. As all agree, the only way the industry can achieve radical improvement is for its members to collaborate. In fact, one of the major exercises that has been undertaken is to benchmark each other's use of subcontractors. There is an intention to extend this effort to other areas. Recent discussion has concerned the use of quality prizes for excellence. These are issues which would never have been envisaged at the outset.

Such work has the effect of providing inspiration. It also allows each of the quality managers to go back to their organisation with renewed enthusiasm and belief. Following one of their meetings, a manager summed up quality in a sentence:

> This TQM is like a sort of religion. Before you hear the word, you think, 'Its got nothing to do with me.' But once you've been hooked, you want to go out and preach the message to others. I find it difficult to understand that it takes others time to appreciate what TQM is. These meetings are vital. I can come and be with the converted, even if it is only for a few hours each month.

It is perhaps not entirely coincidental that Dr Deming was frequently described as a 'prophet unheard' in the West. Yet the Japanese accepted his message with religious fervour. As I have said so often, if proof is required of what quality management can achieve, look at what the Japanese have done.

Other industries have benefited from following the Japanese, and examples can be found in the further reading. However, I would particularly recommend you to a report produced by the CIOB, *Time for real improvement: learning from best practice in Japanese construction R&D* (1995). It describes how the benefits of continuous improvement can be applied to construction. By learning from the Japanese, the 'cycle of decline and low profitability in Britain's construction industry' (ibid.:i) can be halted.

The report contains several conclusions and recommendations. Chief among them is that British construction should learn the lesson from the Japanese, who have put much time and investment into R&D (research and development). This has enabled Japanese construction to 'catch up', and subsequently 'develop a technical capacity which is now extremely high' (ibid.:27).

It stresses the close 'relationship between R&D and the commitment to seeking continuous real improvement' (ibid.:28). As part of the 'quest' for continuous real improvement, it advises that the following features are part of the process (ibid.):

1. Participation is stimulated so that 'everybody feels part of the process, however small their individual contribution may be'.
2. That there is fundamental change to the industry, the result of which is 'deep-seated and continuous transformation'.
3. 'Incremental improvement is brought about gradually by small step-changes'.
4. That there is a search for permanent and continuous improvement.
5. That the effort to finding improvement is holistic, and is applied to all parts of the industry.
6. In what is a reference to benchmarking, that the process is reflective, and is willing to learn from what is already done well, and thus encouraged elsewhere, but also that learning comes from seeing 'what other do even better'.

What Japanese industry has achieved in general and Japanese construction in particular, are aspects of the same phenomenon. My objective throughout has been to demonstrate how the theories of quality are generic. Construction is neither exceptional nor unique nor in any way special. Indeed, given some of its 'traditional' problems, perhaps it needs quality management more than most.

SUMMARY

TQM *is* possible in construction. Contractors and subcontractors can learn from the accomplishments of other industries, many of them achieved by emulating Japan.

The encouragement of cultural change to facilitate TQM will require a different approach to the accepted 'way of the industry'. This will apply from the top (senior management) to the bottom, and it will embrace the subcontractors and suppliers. It will not be easy; much effort will be required. The alternative is to continue to be seen as inefficient and wasteful; surely no one would consider this desirable.

QUESTIONS

Discussion

The lessons of TQM, and the radical improvements it brings, have come from Japan, especially Japanese manufacturing. And many have drawn the conclusion that initiatives such as TQM are only suitable for industries which produce high quality electrical equipment and automobiles. This is patently untrue. But how can an industry such as construction emulate the manufacturing sector and other businesses which use TQM. Agree on five commitments for achieving this objective.

Individual

Write an essay of 2000 words to cover all three items:

(a) Critically examine the concept of cultural change; consider what is needed to achieve it.

(b) Describe two examples of service you have received in the last year or so, one of them inferior and one of them superior. Choose one of the examples and give your opinion of why the service turned out that way.

(c) How can you as an individual contribute to improving the construction industry, and what do you require from others?

FURTHER READING

Bate, P. (1995) *Strategies for cultural change.* Butterworth-Heinemann, Oxford.

Binney, G. (1992) *Making quality work: lessons from Europe's leading companies.* Economist Intelligence Unit, London.

Brown, M. G., D. E. Hitchcock and M. L. Willard (1994) *Why TQM fails, and what to do about it.* Irwin, New York.

Buchanan, D. and D. Boddy (1992) *The expertise of the change agent.* Prentice Hall, Hemel Hempstead.

Carnall, C. A. (ed.) (1997), *Strategic change.* Butterworth-Heinemann, Oxford.

Ferry, J. (1993) *The British renaissance: learn the secrets of how six British companies are conquering the world.* William Heinemann, London.

Joynson, S. and A. Forrester (1995) *Sid's heroes.* BBC Books, London.

Lessem, R. (1994) *Total quality learning: building a learning organization.* Blackwell, Oxford.

McCabe, S. (1996) Making quality work in construction – a study of the effects of its introduction on ten contracting firms, in B. Greenhalgh (ed.) *Practice management for land, construction and property professionals.* E. & F. N. Spon, London, pp. 288–95.

McCabe, S., D. Crooke and J. Rooke (1996) Change management – a consideration of how theory can inform understanding of practice, in *Proceedings of the Twelfth ARCOM Conference,* Sheffield Hallam University, pp. 368–77.

McCabe, S., D. Crook, J. Rooke and D. Seymour (1997) 'Construction's experiences of attempting to become world class: the role of academics assisting practitioners', in *Proceedings of the Thirteenth ARCOM Conference.* King's College, Cambridge, pp. 291–300.

McCabe, S., J. Rooke and D. Seymour (1996) The importance of leadership in change initiatives, in D. A. Langford and A. Retik (eds) *The organization and management of construction: shaping theory and practice,* Vol. 2. E. & F. N. Spon, London, pp. 589–98.

McCabe, S., J. Rooke, D. Seymour and P. Brown (forthcoming) Quality managers, authority and leadership, in *Construction management and economics.* E. & F. N. Spon, London.

McGeorge, D. and A. Palmer (1997) *Construction management, new directions.* Blackwell Science, Oxford.

Oakland, J. S. and L. Porter (1995) *Total quality management: text with cases.* Butterworth-Heinemann, Oxford.

Pettigrew, A. and R. Whipp (1993) *Managing change for competitive success.* Blackwell, Oxford.

Taylor, B. (1994) *Successful change strategies: chief executive in action.* Director Books, Hemel Hempstead.

Ward, M. (1994) *Why your corporate culture change isn't working, and what to do about it.* Gower, Aldershot.

REFERENCES

Adair, J. (1983) *Effective leadership*. Pan, London.

Adams, J. S. (1963) Towards an understanding of inequity. *Journal of Abnormal and Social Psychology*, **67**, 422–36.

Alderfer, C. P. (1972) *Existence, relatedness and growth: human needs in organizational settings.* Free Press, New York.

Argyris, C. (1964) *Integrating the individual and the organization*. John Wiley, New York.

Argyris, C. (1993) *On organizational learning*. Blackwell Business, Cambridge, Massachusetts.

Argyris, C. and D. Schon (1978) *Organizational learning: a theory of action perspective.* Addison-Wesley, Wokingham, England.

Ashford, J. L. (1989) *The management of quality in construction*. E. & F. N. Spon, London.

Avolio, B. J. and B. M. Bass (1985) Transformational leadership, charisma and beyond. *Working paper*, School of Management, State University of New York, Binghamton, New York.

Ball, M. (1988) *Rebuilding construction: economic change in the British construction industry.* Routledge, London.

Bank, J. (1992) *The essence of total quality management*. Prentice Hall, Hemel Hempstead.

Bass, B. M. (1990) From transactional to transformational leadership: learning to share the vision. *Organizational Dynamics*, **Winter**, 19–31.

Bass, B. M. and B. J. Avolio (1990) Developing transformational leadership: 1992 and beyond. *Journal of European Industrial Training*, **Jan**, 23.

BBC (1993) Crazy times, crazy organizations: the Tom Peters seminar. London, November 1993.

Belbin, R. M. (1981) *Management teams: why they succeed or fail*. Butterworth-Heinemann, Oxford.

Bell, D., P. McBride and G. Wilson (1994) *Managing quality*. Butterworth-Heinemann, Oxford.

Benjamin, G. and C. Mabey (1993) Facilitating radical change, in C. Mabey and B. Mayon-White (eds) *Managing change*. Open University/Paul Chapman, London.

Bennis, W. (1984) The 4 competencies of leadership. *Training and Development Journal*, **Aug**, 15–19.

Billingham, E. and A. Stewart, (1993) Standard fireworks, *Building*, 19 November, pp. 20–21.

Binney, G. and C. Williams (1995) *Leading into the future: changing the way that people change organisations*. Nicholas Brealey, London.

Blake, R. R. and A. A. McCanse (1991) *Leadership dilemmas: grid solutions*. Gulf Publishing, Houston TX.

Blake, R. R. and J. S. Mouton (1985) *The managerial grid*. Gulf Publishing, Houston TX.

Blank, W., J. R. Weitzel and S. G. Green (1990) A test of the situational leadership theory. *Personnel Psychology*, **43**, 579–97.

Bounds, G., L. Yorks, M. Adams and G. Ranney (1994) *Beyond total quality management: towards the emerging paradigm*. McGraw-Hill, New York.

Bounds, G. M., G. H. Dobbins and O. S. Fowler (1995) *Management: a total quality perspective*, South-Western College Publishing, Cincinnati OH.

Briggs, M. I. (1987) *Introduction to type: a description of the theory and applications of the Myers Briggs type indicator*. Consulting Psychologists Press, Palo Alto CA.

Brown, A. (1995) *Organisational Culture*. Pitman, London.

Brown, T. (1993) *Understanding BS 5750, and other quality systems*. Gower, Aldershot.

Bryman, A. (ed.) (1988) *Doing research in organizations*. Routledge, London.

BS 4778 Part 2 (1991) *Quality concepts and related definitions*. BSI, London.

BS 6143 (1992) *Guide to the economics of quality-process cost model*. BSI, London.

BS 7850 Part 1 (1992) *Total quality management: guide to management principles*. BSI, London.

BS 7850 Part 2 (1992) *Total quality management: guide to quality improvement methods*. BSI, London.

BS EN ISO 8402 (1995) *Quality management and quality assurance – vocabulary*. (Formerly BS 4778: Part 1, 1987/ISO 8402, 1986). BSI, London.

BS EN ISO 9000 (1994) *Series of standards for the management of quality*. BSI, London.

Building Economic Development Council (1987) *Achieving quality on building sites*. NEDO, London.

Building Employers Confederation (1990) *Quality management for builders*. BEC, London.

Bullock, R. J. and D. Batten (1985) It's just a phase we're going through: a review and synthesis of OD phase analysis. *Group and Organization Studies*, 10 December, pp. 383–412.

Burnes, B. (1996) *Managing change: a strategic approach to organisational dynamics*, 2nd edn. Pitman, London.

Camp, R. C. (1989) *Benchmarking: the search for industry best practices that lead to superior performance*. ASQC Quality Press, Milwaukee WI.

Chartered Institute of Building (1990) *Quality assurance in the building process*. CIOB, Ascot.

Chartered Institute of Building (1995) *Time for real improvement: learning from best practice in Japanese construction R&D*, report of the DTI Overseas Science and Technology Expert Mission to Japan, December 1994. CIOB, Ascot.

Chevin, D. (1991) Never mind the quality. *Building*, 15 November, pp. 48–49.

Clark, K. B. and T. Fujimoto (1988) The European mode of product development: challenge and opportunity. *IMVP working paper*, May 1988.

Clarke, C. and S. Pratt (1985) Leadership's four-part progress. *Management Today*, Mar, 84–86.

Clutterbuck, D. (1994) *The power of empowerment: release the hidden talents of your employees*. BCA and Kogan Page, London.

Conger, J. A. and R. N. Kanungo (1988a) Behavioral dimensions of charismatic leadership, in J. A. Conger and R. N. Kanungo (eds) *Charismatic leadership*. Jossey-Bass, San Francisco CA, p. 79.

Conger, J. A. and R. N. Kanungo (1988b) Training charismatic leadership: a risky and critical task, in J. A. Conger and R. N. Kanungo (eds) *Charismatic leadership*. Jossey-Bass, San Francisco CA, pp. 309–23.

Construction Industry Training Board (1990) *A guide to managing quality in construction*. CITB, Kings Lynn.

Coulson-Thomas, C. (1996) *Business process reengineering: myth and reality*. Kogan Page, London.

Crainer, S. (1996) *Key management ideas: thinking that changed the management world*. Pitman, London.

Crainer, S. (1997) *The ultimate business library: 50 books that made management*. Capstone, Oxford.

Crosby, P. (1979) *Quality is free*. McGraw-Hill, New York.

Cross, R. and P. Leonard (1994) Benchmarking: a strategic and tactical perspective, in B. G. Dale (ed.), *Managing quality*. Prentice Hall, Hemel Hempstead, pp. 497–513.

Cummings, T. G. and E. F. Huse (1989) *Organization development and change*. West, St Paul MN.

Dale, B. G. (1994a) *Managing quality*. Prentice Hall, Hemel Hempstead.

Dale, B. G. (1994b) Quality management systems, in B. G. Dale (ed.) *Managing quality*. Prentice Hall, Hemel Hempstead, pp. 333–61.

Dale, B. G. (1994c) Quality costing, in B. G. Dale (ed.) *Managing quality*. Prentice Hall, Hemel Hempstead, pp. 209–30.

Dale, B. G. and R. J. Boaden (1994) The use of teams in quality improvement, in B. G. Dale (ed.) *Managing quality*. Prentice Hall, Hemel Hempstead, pp. 514–30.

Dale, B. G. and C. Cooper (1992) *Total quality and human resources: an executive guide*. Blackwell, Oxford.

Dale, B. G. and J. J. Plunkett (1990) *The case for costing quality*. DTI, London.

Dale, B. G. and J. J. Plunkett (1991) *Quality costing*. Chapman & Hall, London.

Dale, B. G., R. J. Boaden and D. M. Lascelles (1994) Total quality management: an overview, in B. G. Dale (ed.) *Managing quality*. Prentice Hall, Hemel Hempstead, pp. 3–40.

Dale, B. G., H. S. Bunney and P. S. Shaw (1994) Quality management tools and techniques: an overview, in B. G. Dale (ed.) *Managing quality*. Prentice Hall, Hemel Hempstead, pp. 379–410.

Dale, B. G., D. M. Lascelles and R. J . Boaden (1994) Levels of total quality management adoption, in B. G. Dale (ed.) *Managing quality*. Prentice Hall, Hemel Hempstead, pp. 117–27.

Dale, B. G., D. M. Lascelles and A. Lloyd (1994) Supply chain management and development, in B. G. Dale (ed.) *Managing quality*, Prentice Hall, Hemel Hempstead, pp. 292–315.

de Charms, R. (1968) *Personal causation: the internal affective determinants of behaviour*. Academic Press, New York.

Deci, E. L. (1975) *Intrinsic motivation*. Plenum, New York.

Deci, E. L. (1976) The hidden costs of rewards. *Organisational Dynamics*, **4**, 61–72.

Deming, W. E. (1986) *Out of the crisis*. MIT Press, Boston MA.

Dobson, P. (1988) Changing culture. *Employment Gazette*, **Dec**, 647–650.

Drennan, D. (1992) *Transforming company culture*. McGraw-Hill, London.

Drew, H. E. (1972) Defence quality assurance in the United Kingdom. *Quality Progress*, **5**, 28–32.

Drucker, P. (1946) *The concept of the corporation*. Mentor, New York (reprinted).

Duncan, J., B. Thorpe and P. Sumner (1990) *Quality assurance in construction*. Gower, Aldershot.

Dunphy, D. D. and D. A. Stace (1993) The strategic management of corporate change. *Human Relations*, **46**, 905–18.

Easterby-Smith, M., R. Thorpe and A. Lowe (1991) *Management research: an introduction*. Sage, London.

Eldridge, J. E. T. and A. D. Crombie (1974) *A sociology of organisations*. Allen & Unwin, London.

Feigenbaum, A. V. (1956) Total quality control. *Harvard Business Review*, **34**, 93–101.

Feigenbaum, A. V. (1961) *Total quality control: engineering and management*. McGraw-Hill, New York.

Fiedler, F. E. (1976) *A theory of leadership effectiveness*. McGraw-Hill, New York.

Filley, A. C., R. J. House and S. Kerr (1976) *Managerial process and organisational behavior*, 2nd edn. Scott Foresman, New York.

Fleishman, E. A. (1974) Leadership climate, human relations training and supervisory behavior, in E. A. Fleishman and A. R. Bass (eds) *Studies in personnel and industrial psychology*, 3rd edn. Dorsey, New York.

Flood, R. L. (1993) *Beyond TQM*. John Wiley, Chichester.

Foster, A. (1989) *Quality assurance in the construction industry*. Hutchinson, London.

French, W. L. and C. H. Bell (1984) *Organisation development*. Prentice Hall, Englewood Cliffs NJ.

Garvin, D. A. (1993) Building a learning organization. *Harvard Business Review*, July/Aug, 78–91.

George, C. S. (1972) *The history of management thought*. Prentice Hall, Englewood Cliffs NJ.

Gilbert, J. (1992) *How to eat an elephant: a slice by slice guide to total quality management*. Tudor, Merseyside.

Griffith, A. (1990) *Quality assurance in building*. Macmillan, London.

Hackman, J. R. and E. E. Lawler (1971) Employee reactions to job characteristics. *Journal of Applied Psychology*, **55**, 259–68.

Hackman, J. R. and G. R. Oldham (1980) *Work Redesign*, Addison Wesley Publishing Inc, Reading, MA, USA.

Hamel, G. and C. K. Prahalad (1994) *Competing for the future*. Harvard Business School Press, Boston MA.

Hamilton, N. (1991) The Great Divide. *New Builder*, 18 April, p. 20.

Handy, C. (1976) *Understanding organizations*. Penguin, London.

Handy, C. (1978) *Gods of management*. Arrow Business Books, London.

Handy, C. (1984) *The future of work*. Blackwell, Oxford.

Handy, C. (1989) *The Age of Unreason*. Arrow Business Books, London.

Handy, C. (1990) *Inside organizations: 21 ideas for managers*. BBC Books, London.

Handy, C. (1991) *Waiting for the mountain to move and other reflections on life*. Arrow Business Books, London.

Handy, C. (1994) *The empty raincoat*. Hutchinson, London.

Handy, C. (1995) *Beyond certainty: the Changing world of organizations*. Century Books, London.

Handy, C. and J. Constable (1988) *The making of managers*. Longman, London.

Harari, O. (1993) Ten reasons why TQM doesn't work. *Management Review*, Jan, 33–38.

Harrington, H. J. (1991) *Business process improvement: the breakthrough strategy for total quality, productivity, and competitiveness*. McGraw-Hill, New York.

Harvey, R. C. and A. Ashworth (1993) *The construction industry of Great Britain*. Butterworth-Heinemann, Oxford.

Hater, J. J. and B. M. Bass (1988) Supervisors' evaluation and subordinates' perceptions of transformational and transactional leadership. *Journal of Applied Psychology*, Nov, 695–702.

Hersey, P. and K. H. Blanchard (1974) So you want to know your leadership style? *Training and Development Journal*, Feb, 1–15.

Hersey, P. and K. H. Blanchard (1982) *Management of Organization Behavior: utilizing human resources*, 4th edn., Prentice Hall, Englewood Cliffs, NJ.

Herzberg, F. (1968) One more time: how do you motivate employees? *Harvard Business Review*, **46**, 53–62.

Hillebrandt, P. M. (1984) *Analysis of the British construction industry*. Macmillan, London.

House, R. J. (1971) A path–goal theory of leader effectiveness. *Administrative Science Quarterly*, Sept, 321–28.

House, R. J. (1977) A 1976 theory of charismatic leadership, in J. G. Hunt and L. L. Larson (eds) *Leadership: the cutting edge*. Southern Illinois Press, Carbondale IL.

House, R. J., J. Woycke and E. M. Fodor (1988) Charismatic and non-charismatic leaders: differences in behavior and effectiveness, in J. A. Conger and R. N. Kanungo (eds) *Charismatic leadership*. Jossey-Bass, San Francisco CA, pp. 103–4.

Howell, J. M. and P. J. Frost (1989) A laboratory study of charismatic leadership. *Organisational Behavior and Human Decision Processes*, **April**, 243–69.

Huczynski, A. A. (1996) *Management gurus: what makes them and how to become one*. International Thomson Business Press, London.

Hughes, T. and T. Williams (1991) *Quality assurance: a framework to build on*. BSP Professional Books, Oxford.

Imai, M. (1986) Kaizen: the key to Japan's competitive success. Random House, New York.

Ishikawa, K. (1985) *What is total quality control? The Japanese way* (translated by D. J. Lu). Prentice Hall, Englewood Cliffs NJ.

Jackson, P. and D. Ashton (1995a) *Achieving BS EN ISO 9000*. Kogan Page, London.

Jackson, P. and D. Ashton (1995b) *Managing a quality system using BS EN ISO 9000 (formerly BS 5750)*. Kogan Page, London.

Jennings, E. E. (1961) The anatomy of leadership. *Management of Personnel Quarterly*, **1**, 2.

Joynson, S. and A. Forrester (1995) *Sid's heroes*. BBC Books, London.

Juran, J. M. (1950) *Quality control handbook*. McGraw-Hill, New York (reprinted 1982).

Kahn, R. and Katz (1960) Leadership practices in relation to productivity and morale, in D. Cartwright and A. Zander (eds) *Group dynamics: research and theory*, 2nd edn., Row Patterson, Elmsford, New York.

Kanter, R. M. (1984) *The change masters: corporate entrepreneurs at work*. Allen & Unwin, London.

Kanter, R. M. (1989) *When giants learn to dance: mastering the challenges of strategy, management and careers in the 1990s*. Simon and Schuster, London.

Kanter, R. M., B. A. Stein and T. D. Jick (1992) *The challenge of organizational change*. Free Press, New York.

Karlof, B. and S. Ostblom (1993) *Benchmarking: a Signpost to excellence in quality and productivity*. John Wiley, Chichester.

Kearns, D. T. and D. A. Nadler (1992) *Prophets in the dark: how Xerox reinvested itself and beat back the Japanese*. Harper, New York.

Kennedy, C. (1994) *Managing with the gurus*. Century, London.

Kirkpatrick, A. and E. A. Locke (1991) Leadership: do traits really matter? *Academy of Management Executive*, **May**, 48–60.

Kormanski, C. and A. Mozenter (1987) A new model for team building: a technology for today and tomorrow, in *The 1985 Annual: Developing Human Resources*. University Associates, San Diego CA.

Kotter, J. P. (1990) What leaders really do. *Harvard Business Review*, **May/June**, 103.

Kotter, J. P. and J. L. Heskett (1992) *Culture and performance*. Free Press, New York.

Kramer. H. (1975) The philosophical foundations of management rediscovered. *Management International Review*, **15**, 47–55.

Krech, D., R. S. Crutchfield and E. L. Ballachey (1962) *Individual in society*. McGraw-Hill, New York.

Kuhn, T. S. (1962) *The structure of scientific revolutions*. University of Chicago Press, Chicago IL.

Lascelles, D. M. and B. G. Dale (1993) *The road to quality*. IFS Publications, Bedford.

Latham, M. (1994) *Constructing the team*. HMSO, London.

Lewin, K. (1958) Group decisions and social change, in G. E. Swanson, T. M. Newcomb and E. L. Hartley (eds) *Readings in social psychology*. Holt, Reinhart and Winston, New York.

Locke, E. A. (1968) Towards a theory of task motivation and extrinsic reward upon high intrinsic motivation. *Organizational Behavior and Human Performance*, 17, 275–88.

Logothetis, N. (1992) *Managing for total quality: from Deming to Taguchi and SPC*. Prentice Hall, Hemel Hempstead.

Macbeth, D. K. and N. Ferguson (1994) *Partnership sourcing: an integrated supply chain management approach*. Pitman, London.

McBride, J. and N. Clark (1996) *20 steps to better management*. BBC Books, London.

McCabe, S. (1997) Using suitable tools for researching what quality managers in construction organisations actually do. *Journal of Construction Procurement*, Volume 3, No. 2, pp. 72–87.

McClelland, D. C. (1961) *The Achieving Society*, Van Norstrand, New York.

McGregor, D. (1960) *The human side of enterprise*. McGraw-Hill, New York.

Machan, D. (1989) The charisma merchants. *Forbes*, 23 January, pp. 100–101.

Manson, M. M. and B. G. Dale (1989) The operating characteristics of quality circles and yield improvement teams: a case study comparison. *European Management Journal*, 7, 287–85.

Maslow, A. H. (1943) A theory of human motivation. *Psychology Review*, 50, 370–96.

Maslow, A. H. (1954) *Motivation and personality*, Harper and Row, New York.

Mayo. A. (1993) Learning at all organizational levels, in P. Sadler (ed.) *Learning more about learning organisations*. AMED, London.

Mayo, A. and Lank, E. (1994) *The power of learning: a guide to gaining competitive advantage*. Institute of Personnel Development, London.

Micklethwait, J. and A. Wooldridge (1996) *The witch doctors: what the management gurus are saying, why it matters, and how to make sense of it*. Heinemann, London.

Morrison, S. J. (1994) Managing quality: an historical review, in B. G. Dale (ed.) *Managing quality*. Prentice Hall, Hemel Hempstead, pp. 41–79.

Morton, C. (1994) *Becoming world class*. Macmillan, Basingstoke.

Mullins, L. J. (1993) *Management and organisational behaviour*, 3rd edn. Pitman, London.

Munro-Faure, L., M. Munro-Faure and E. Bones (1993) *Achieving quality standards: a step-by-step guide to BS 5750/ISO 9000*. Pitman, London.

Nicholls, J. R. (1985) A new approach to situational leadership. *Leadership and Organization Development Journal*, 6, 2–7.

Nonaka, I. (1988) Creating organizational order out of chaos: self-renewal in Japanese firms. *Harvard Business Review*, Nov/Dec, 96–104.

Oakland, J. S. (1993) *Total quality management: the route to improving performance*, 2nd edn. Butterworth-Heinemann, Oxford.

Oliver, N. and B. Wilkinson (1992) *The Japanization of British industry: new developments in the 1990s*, 2nd edn. Blackwell, Oxford.

Ouchi, W. G. (1981) *Theory Z: how American business can meet the Japanese challenge*. Addison-Wesley, Reading MA.

Parker, D. (1993) Not perfect. *New Builder*, 17 September, p. 26.

Pascale, R. T. and Athos, A. G. (1982) *The art of Japanese management*. Penguin, London.

Pateman, J. (1986) There's more to quality than quality assurance. *Building Technology and Management*, Aug/Sept, 16–18.

Pedler, M., J. Burgoyne and T. Boydell (1991) *The Learning Company: a strategy for sustainable development*. McGraw Hill, London.

Peppard, J. and P. Rowland (1995) *The essence of business process re-engineering*. Prentice Hall, Hemel Hempstead.

Peters, T. (1989) *Thriving on chaos*. Macmillan, London.

Peters, T. (1992) *Liberation management.* Macmillan, London.

Peters, T. (1994a) *The Tom Peters seminar: crazy times call for crazy organizations.* Macmillan, London, and Vintage Books, New York.

Peters, T. (1994b) *The pursuit of wow! Every person's guide to topsy turvy times.* Macmillan, London, and Vintage Books, New York.

Peters, T. and N. Austin (1985) *A passion for excellence.* Collins, London.

Peters, T. and Waterman, R. (1982) *In search of excellence: lessons from America's best-run companies.* Harper & Row, New York.

Pettigrew, A. and R. Whipp (1993) Understanding the environment, in C. Mabey and B. Mayon-White (eds) *Managing change,* 2nd edn. Open University/Paul Chapman, London.

Philips, N. V. (1985) *Co-makership: purchasing in Benelux.* Philips, Eindhoven.

Price, F. (1990) The quality concept and objectives, in D. Lock (ed.) *Gower handbook of quality management.* Gower, Aldershot, pp. 3–11.

Rachlin, H. (1970) *Modern behaviorism.* W. H. Freeman, New York.

Rice, R. W. (1978) Psychometric properties of the esteem for the least preferred co-worker (LPC) scale. *Academy of Management Review,* **Jan,** 106–18.

Ridout, G. (1994) Standard fireworks. *Building,* 2 December, pp. 30–32.

Robbins, S. (1994) *Management,* 4th edn. Prentice Hall, Englewood Cliffs NJ.

Rodrigues, C. A. (1988) Identifying the right leader for the right situation. *Personnel,* **Sept,** 43–46.

Sadler, P. (1995) *Managing change.* Kogan Page, London.

Salaman, G. (1979) *Work organisations.* Longman, London.

Scarborough, H. and J. M. Corbett (1992) *Technology and organization: power, meaning and design,* Routledge, London.

Schein, E. H. (1985) *Organizational culture and leadership: a dynamic view.* Jossey-Bass, San Francisco, CA.

Schein, V. E. (1985) Organisation realities: the politics of change. *Training and Development Journal,* **Feb,** 39–40.

Schonberger, R. (1990) *Building a chain of customers.* Free Press, New York.

Semler, R. (1993) *Maverick!* Century, London.

Senge, P. (1990) *The fifth discipline: the art and practice of the learning organization.* Doubleday, New York.

Senge, P., A. Kleiner, C. Roberts, R. B. Ross and B. J. Smith (1994) *The fifth discipline fieldbook: strategies and tools for building a learning organization.* Nicholas Brealey, London.

Shaffir, W. B. and R. A. Stebbins (eds) (1991) *Experiencing fieldwork: an inside view of qualitative research.* Sage, London.

Shewhart, W. (1931) *Economic control of quality of manufactured product.* Van Norstrand, New York.

Sirkin, H. L. (1993) The employee empowerment scam. *Industry Week,* 18 October, p. 58.

Skinner, B. F. (1953) *Science and human behavior.* Macmillan, New York.

Stahl, M. J. (1995) *Management: total quality in a global environment.* Blackwell, Oxford.

Tannenbaum, R. and W. H. Schmidt (1973) How to choose a leadership pattern. *Harvard Business Review,* **May/June,** 162–75, 178–80.

Taylor, F. W. (1911) *The principles of scientific management.* Harper, New York (1947 edition).

Taylor, M. and H. H. Hosker (1992) *Quality assurance for building design.* Longman, Harlow.

Thomas, A. B. (1993). *Controversies in management.* Routledge, London.

Thomas, B. (1995) *The human dimension of quality.* McGraw-Hill, Maidenhead.

Tuckman, B. W. and M. A. Jensen (1977) Stages of small group development revisited. *Group and Organizational Studies,* **2,** 419–27.

Uttal, B. (1983) The corporate culture vultures. *Fortune*, 17 October, pp. 66–72.

Vecchio, R. P. (1987) Situational leadership theory: an examination of a prescriptive theory. *Journal of Applied Psychology*, **72**, 444–52.

Vroom, V. H. (1964) *Work and motivation*. John Wiley, New York.

Vroom, V. H. and P. W. Yetton (1973) *Leadership and decision-making*. University of Pittsburgh Press, Pittsburgh PA.

Walker, R. (1992) Rank Xerox – management revolution. *Long-Range Planning*, Vol. **25**, No. 1, Feb., pp. 9–21.

Walsh, K. (1995) Quality through markets, the new public service management, in A. Wilkinson and H. Willmott (eds) *Making quality critical: new perspectives on organizational change*. Routledge, London, pp. 82–104.

Walton, M. (1989) *The Deming management method*. Mercury Business Books, London.

Warner, F. (1977) *Standards and specifications in the engineering industries*. NEDO, London.

Watson, T. J. (1986) *Management, organisation and employment strategy: new directions in theory and practice*. Routledge, London.

Watson, T. J. (1994) *In search of management: culture, chaos and control in managerial work*. Routledge, London.

Wickens, P. (1987) *The road to Nissan: flexibility, quality, teamwork*. Macmillan, Basingstoke.

Wilkinson, A. (1994) Managing human resources for quality, in B. G. Dale (ed.) *Managing quality*. Prentice Hall, Hemel Hempstead, pp. 273–89.

Wilkinson, A. and H. Wilmott (1995) *Making quality critical: new perspectives on organizational change*. Routledge, London.

Wilton, P. S. (1994) *The quality system development handbook (with ISO 9000)*. Prentice Hall, Hemel Hempstead.

Womack, J. P. and D. T. Jones (1996) *Lean thinking: banish waste and create wealth in your Corporation*. Simon and Schuster, New York.

Womack, J. P., D. T. Jones and D. Roos (1990) *The machine that changed the world*. Rawson Associates, New York.

INDEX

'A concrete story' 204
Adams, J. S. 95
Adair, J. 106–7
Alderfer, C. P. (ERG theory) 93
AQAP (Allied Quality Assurance Publication) 22
Argyris, C. 96, 153
Argyris, C. and D. Schon 168
  Organisational Learning 168
  single/double-loop learning 168
  espoused theories and theories-in-use 168
Ashford, J. L. 19, 186
Avolio, B. J. and B. M. Bass 120

Ball, M. 2, 194
Bass, B. 119
Bass, B. M. and B. J. Avolio 120
Benjamin, G. and C. Mabey 86
Baldridge Award 178
Bank, J. 22, 149
BEDC (Builing Economic Development Council) 2
Beens, L. L. 151
Bell Telephones 31
Bell, D., P. McBride and G. Wilson 30, 105, 154
Benchmarking 4, 148–53
  reasons for 149
  internal benchmarking 149
  competitive 150
  functional/generic 150
Bennis, W. 118
Billingham, E. and A. Stewart 187
Bill's story 184–5
Binney, G. and C. Williams 173
  Keaning into the future 173
Blake, R. R. and A. A. McCanse 107
Blake, R. R. and J. S. Mouton 107
Blank, W., J. R. Weitzel and S. G. Green 113
Bounds, G. M., G. H. Dobbins and O. S. Fowler 94, 98, 108, 119
Bounds, G. M., L. Yorks, M. Adams and G. Ranney 161, 169
  Beyond TQM, towards the emerging paradigm 161

old and new paradigms for customer value strategy 162
old and new paradigms for organisational systems 163
old and new paradigms for continuous improvement 163
BPR (Business process re-engineering) 4, 154–9
  creating the right environment 155
  analyse, diagnose and redesign the processes 157
  restructure the organisation 157
  pilot and roll-out 157
  realise the strategy 157
Brainstorming 126
British Leyland 23
British Quality Award 23, 181
Brown, T. 56, 80
Bryman, A. 187
BS 4891 (1971) *A guide to quality assurance* 22
BS 5750 (1979) *Quality Systems* 22, 186
BS 4778 Part 2 (1991) *Quality concepts and related definitions* 7
BS 6143 (1992), *Guide to economics of quality-process cost model* 132
BS EN ISO 8402 (1995), *Quality management and quality assurance – vocabulary* 2, 49
BS EN ISO 9000 (1994) *Standard for Quality Management* 3, 49
BS 7850 (1992) *Total quality management, part 1, guide to management principles* 12
BS 7850 (1992) *Total quality management, part 2, guide to quality improvement methods* 12
Building Employers Confederation 186
Bullock, R. J. and D. Batten 85
Burnes, B. 24, 81, 84, 86, 87, 90, 164

Camp, R. C. 151
Cause and effect analysis 125–6
CBI (Confederation of British Industry) 22
CEDAC (cause and effect diagrams with additional cards) 126
Checklists 124

Checksheets 124
Chevin, D. 186, 187
CIOB (The Chartered Institute of Building) 1, 5, 186
   Time for real improvement: learning from best practice in Japanese construction R&D 215
Clark, K. B. and T. Fujimoto 170
Clarke, C. and S. Pratt 117
Clutterbuck, D. 98, 101–2
Construction industry training board 186
Communication 10
Conger, J. and R. Kanungo (charismatic leadership) 118, 199
Control charts 41
Conway, W. 3, 31, 46–7
   Nuasha Corporation 46
   five points 47
Coulson-Thomas, C. 154, 158
Coverdale organisation 98
Crainer, S. 35, 44, 81, 166, 167
Crosby, P. 3, 37–9
   ITT 37
   symptoms of 'sick' organisations 38
   'vaccination serum' 38
   fourteen points 39
Cross, R. and P. Leonard 150, 153
Culture 3
Culture change for TQM 79–88
   facilitating change 83, 141
   planned 84
   phases of 85
   emergent 84
   action research 85
   three-step model 85
Cultural change in construction 213–14
Cummings, T. G. and E. F. Huse 84, 85
Customer 4, 13

Dale, B. G. 132
Dale, B. G., R. J. Boaden and D. M. Lascelles 8, 9, 13, 35, 80
Dale, B. G., D. M. Lascelles and R. J. Boaden 175
Dale, B. G., D. M. Lascelles and A. Lloyd 143, 145
Dale, B. G. and C. Cooper 97, 105, 119
Dale, B. G. and J. Plunkett 132
Dantotsu (benchmarking) 148
DEF STAN (Defence Standards) 21
Deming, W. Edwards 2, 20, 29–35, 90
   prize 31
   flow diagram 32
   Out of the Crisis 32
   Deming Cycle 32
   fourteen points 33
Dobson, P. 84
Drennan, D. 81
Drew, H. E. 17
Drucker, P. 98
   The concept of the corporation 99

Duncan, J., B. Thorpe and P. Sumner 186
Dunphy, D. D. and D. A. Stace 86

Easterby-Smith, M., R. Thorpe and A. Lowe 187
Eldridge, J. E. T. and A. D. Crombie 81
Empowerment 3, 98
European quality award 4, 178–81

Factory system 17
FMECA (Failure mode, effect and criticality analysis) 129
Feigenbaum, A. V. 3, 40–1
Fiedler, F. 110–13
Financial Times 6
Fishbone diagrams 41
Fleishman, E. A. (Ohio State studies) 107
Flood, R. L. 42
flowcharts 124
Ford Cars 170
Foster, A. 186
French, W. L. and C. H. Bell 85

Garvin, D. A. 86
'Getting the subbies involved' 193–7
Getting close to the client 208–9
George, C. S. 17
Griffith, A. 186
Gurus of quality 27–48

Hackman, R. and G. Oldham 97
Hamel, G. and C. K. Prahalad 182
   Competing for the future 182
Hamilton, N. 187
Handy, C. 83, 105
   types of culture 83
   Understanding organizations 99
   Gods of management 99
   The future of work 99
   The making of manager 99
   The age of unreason 99
   Inside organisations 99
   Waiting for the mountain to move 99
   The empty raincoat 99
   Beyond certainty 99
Harari, O. 161
Harvey, R. C. and A. Ashworth 2, 194
Hater, J. J. and B. M. Bass 119
Hawthorne experiments 2, 19, 20
Hersey, P. and K. Blanchard (leadership situational theory) 111–14
Herzberg, F. (two-factor theory) 91
Hillebrandt, P. M. 194
Histograms 41
House, R. J. (path-goal theory) 114–15, 118
House, R. J., J. Woycke and E. M. Fodor 119
Howell, J. M. and P. J. Frost 119
Human relations approach 20
Huczynski, A. A. 28
Hughes, T. and T. Williams 50, 186–7

Imai, M. 3, 39–40
Improvement methods 141–60
Improvement teams 135
Industrial Revolution 17
Ishikawa, K. 3, 41–2, 125

Jackson, P. and D. Ashton 56, 65, 70, 80
Japan 1, 2, 17
  characteristics of Japanese organisations 24
  strategy of organisations 25, 81
Jennings, E. E. 106
Jidohka (automation) 89, 169
Joyson, S. and A. Forrester 17, 40
Juran, J. 2, 20, 35–7, 90
  ten points 35
  Juran trilogy 36
JUSE (Union of Japanese Scientists and
    Engineers) 31
Just-in-time 4

Kaizen 39
Kanban 169
Kanter, R. M. 99
  The change masters 99
  When giants learn to dance 99
Kanter, R. M., B. A. Stein and T. D. Jick 82
Kahn, R. and D. Katz (University of Michigan
    studies) 107
Karlof, B. and S. Ostblom 151, 153
Kei-dan-ren (Japanese Chief Executives) 31
Kearns, D. T. (Xerox Corporation) 120, 150
Kearns, D. T. and D. A. Nadler 120
Kennedy, C 16, 20, 43
Kirkpatrick, A. and E. A. Locke (qualities or traits
    of leaders) 105
Kormonski, C. and A. Mozenter 137
Kotter, J. P. 106
Kotter, J. P. and J. L. Heskett 82
Kramer, H. 28
Krech, D., R. S. Crutchfield and E. L. Ballachey
    106
Kuhn, T. 162

Lascelles, D. M. and B. G. Dale 4, 175
  The road to quality 203
Latham report (1994) 1
Leadership 3
  charismatic 3, 117–19
  transformational 3, 119–20
  for TQM 104–21
  functional or group 106
  behavioural theories 107
  contingency theories 110–16
Lean production 4, 169–74
  leadership 171
  teamwork 171
  communication 171
  simultaneous development 171
Lean principles 171

Learning (direct and indirect) 164
Learning organisation 4, 164–8
Lewin, K. 85
Locke, E. A. (goal-setting theory) 95
Logothetis 29

MacArthur, General D. 31
Macbeth, D. K. and N. Ferguson 142
  adversarial relationships 142
  collaborative relationships 143
  five stages of relationship improvement 144
  relationship positioning tool 146
Machan, D. 119
'Maintaining momentum' 198–200
Manson, M. M. and B. G. Dale 135–6
Marks and Spencer 142
Maslow, A. H. (hierarchy of needs) 90
Mayo, A. 167
Mayo, A. and E. Lank 168
  *The power of learning: a guide to competitive
    advantage* 168
Mayo, E. 20
McBride, J. and N. Clark 98, 100
McCabe, S. 187
McCelland, D. C. (socially acquired needs theory)
    93
McGregor, D. 46
  theory X and Y 46, 90
Micklethwaite, J. and A. Wooldridge 29, 173
Morrison, S. J. 22
Morton, C. 169
Motivation 3, 89–102
  content theories 92–3
  process theories 93–6
MoT (moment of truth) 129
Muda 4
Mullins, L. J. 109
Munro-Faure, L., M. Munro-Faure and E. Bones
    56

National quality campaign 186
NATO (North Atlantic Treaty Organisation)
    21
NACCB (National Accreditation Council for
    Certification Bodies) 73
Nicholls, J. R. 115
Nominal group technique 127
Nonaka, I. 86

Oakland, J. 122–3, 133, 148, 178
Ohno, T. 3, 42, 170
'Old faces for a new job' 188–91
Oliver, N. and B. Wilkinson 143
  The Japanization of British industry 143
Organisational culture 81
Ouchi, W. G. 3, 45–6
  theory Z 45, 96–7
  seven characteristics of industrial decline in
    American organisations 46

Pareto analysis 3, 41, 124–5
Parker, D. 187
Partnering 4, 142–8
Pascale, R. T. and A. G. Athos 3, 45, 81
Pateman, J. 186
Pedler, M., J. Burgoyne and T. Boydell 164
Peppard, J. and P. Rowland 154
People, and TQM 89–103
Peters, T. 82, 87
  Thriving on chaos 99
  Liberation management 99
  The Tom Peters seminar 99
  The pursuit of wow! 99
Peters, T. and N. Austin 99
  A passion for excellence 99
Peters, T. and R. Waterman 3, 43–5
  In Search of Excellence 43, 81, 99
  eight attributes of excellent organisations
    43
  McKinsey 7-S framework 44
Pettigrew, A. and R. Whipp 87, 117, 164
Philips, N. V. 142
Poka-yoke 89
Price, F. 1
Process analysis 57
Project teams 134
PSA (Property services agency) 16, 186

Quality
  inspection 7
  control 8
  inspectors 18
QA (Quality Assurance) 2, 8, 49–62
  benefits 9
  tender lists 11
  right first time 11
  pitfalls of 12
  cost of 12
  third party assessment 50
  management responsibility (clause 4.1) 52
  quality system (clause 4.2) 53
  contract review (clause 4.3) 53
  design control (clause 4.4) 54
  document and data control (4.5) 54
  purchasing (clause 4.6) 55
  contol of customer-supplied (clause 4.7) 55
  product identification and traceability
    (clause 4.8) 56
  process control (clause 4.9) 56
  inspection and testing (clause 4.10) 57
  control of inspection, measuring and test
    equipment (clause 4.11) 58
  control of non-conforming products
    (clause 4.13) 58
  corrective action and preventative action
    (clause 4.14) 59
  handling, storage, packaging, preservation and
    delivery (clause 4.15) 59
  control of quality records (clause 4.16) 59

  internal quality audits (clause 4.17) 60
  training (clause 4.18) 60
  servicing (clause 4.19) 60
  statistical techniques (clause 4.20) 61
Quality circles 134
Quality costing 132–3
QFD (Quality function deployment) 127–9
  house of quality 127–8
Quality managers 63–78
  administration skills 65
  management skills 65
  writing procedures interactively 67, 191–3
  maintaining momentum 68
  staff training 69
  quality system documentation control 71
  monitoring performance 71
  audits 72
  working with assessors 73
  sorting out problems 74
  changing the system 76
  'as enforcer' 191
  'as charismatic' 193

Rachlin, H. 94
Rank Xerox 150–3
Reinforcement theory 93
Ridout, G. 187
Rodrigues, C. A. 117
Roles within a team 137–9
  Belbin, M., *Management teams, why they succeed
    or fail* 137–8
  Myers-Briggs type indicator 138–9
  Oakland's DRIVE model 139
Rover cars
  Longbridge plant 23
Robbins, S. 13, 98, 117

Sadler, P. 81, 142, 148, 154, 164
Salaman, G. 84
Scarbrough, H. and J. M. Corbett 142
Scatter diagrams 41, 124
Schein, E. H. 84
Schein, V. E. 158
Schonberger, R. 3, 47
  Building a Chain of Customers 47
  twelve dimensions of quality 47
Scientific management 18
Semler, R. 99
  Maverick 99
Senge, P. 153, 167
  The fifth discipline 166
  systems thinking 166
  personal mastery 166
  mental models 166
  shared vision 166
  team leaning 166
Senior managers and QA 197–8
Senior managers and TQM 211–13
Setting up a quality circle onsite 205–8

Seven tools of quality   130
Shaffir, W. B. and R. A. Stebbins   187
Shewhart, W.   20, 29, 90
Shingo, S.   3, 42–3
   kanban   42
   poka-yoke   42
Shojinka (flexible working)   169
Sirkin, H. L.   100
Skinner, B. F.   94
Soikufu (creative thinking)   169
SPC (Statistical process control)   3, 20, 29, 131–2
   variation   30
   uncontrolled and controlled   30
   causes of, special and common   30

Stahl, M. J.   116, 118
Stratification   41

Tally charts   41, 124
Tannenbaum, R. and W. H. Schmidt (continuum
   of leadership)   109
Taylor, F. W.   18
Taylor, M. and H. H. Hosker   17
Teamworking   4, 133–9
Thomas, A. B.   45, 105
Thomas, B.   125, 212
Toyota   4, 42, 169
Toyoda, E.   169
TQM (Total Quality Management)   2, 12

TQM, life after   161–83
Tuckman, B. W. and M. A. Jensen   136–7

Using subcontractor expertise   209–11
Uttal, B.   84

Vecchio, R. P.   113
Vroom, V. (expectancy theory)   94
Vroom, V. and P. Yetton (leadership participation
   model)   116

Walton, M.   31
Walker, W.   120
Walsh, K.   186
Watson, T.   82, 187
Warner report   51
Wickens, P.   23
Wilkinson, A. and H. Willmott   43, 122
Wilton, P. S.   54
Womack, J. P. and D. T. Jones   172
   Lean thinking   172
   specify value   172
   identify value stream   172
   flow, pull and perfection   173
Womack, J. P., D. T. Jones and D. Roos   143,
   170–2
   The machine that changed the world   143, 169
World class   4, 175–81
   stages towards   175–81